Connecting Algebra and Geometry through Technology

Applying Geometry Expressions in the Algebra 2 and Pre-Calculus Classrooms

Jim Wiechmann
Tualatin High School, Tualatin, OR, USA

Saltire Software, Inc.
Tigard, OR, USA
www.saltire.com
www.geometryexpressions.com

Saltire Software
P.O. Box 230755
Tigard, OR 97281-0755
http://www.geometryexpressions.com/
http://www.saltire.com/
support@saltire.com

Table of Contents

Introduction

The National Council of Teachers of Mathematics (NCTM) Vision for School Mathematics invites us to *"Imagine a classroom, a school, or a school district where all students have access to high-quality, engaging mathematics instruction."* It goes on to describe how this may take place in the classroom:

> *Teachers help students make, refine, and explore conjectures on the basis of evidence and use a variety of reasoning and proof techniques to confirm or disprove those conjectures. Students are flexible and resourceful problem solvers. Alone or in groups and with access to technology, they work productively and reflectively, with the skilled guidance of their teachers. Orally and in writing, students communicate their ideas and results effectively. They value mathematics and engage actively in learning it.* (National Council of Teachers of Mathematics (NCTM). *Principals and Standards for School Mathematics.* Reston, VA: NCTM, 2000).

Our goal in writing this book is to provide examples of how a symbolic geometry system, Geometry Expressions, can begin to make this happen. Geometry Expressions provides a playground where students can discover their own mathematics. They will begin to see mathematics as something that is created, not just a set of facts made up long ago. Once students take ownership of their mathematics, they will be more apt to "work productively and reflectively, with the skilled guidance of their teachers."

The graphical, interactive nature of Geometry Expressions brings life into a field that might otherwise seem irrelevant. The symbolics embedded in Geometry Expressions offer an algebraic view of the mathematics in concert with a geometric view, blurring the artificial line between the two. The smooth interface between Geometry Expressions and Computer Algebra Systems (CAS) adds another powerful resource for solving problems. These technologies can work together to change the way mathematics is done, in the same way that technology has changed the way architectural design is done; with computers managing the details while humans create the grand vision.

The units presented in this book are a jumping-off point for using Geometry Expressions in the classroom. Use the units to gauge the potential of this powerful software, and as a guide to applying Geometry Expressions in your own classroom. We trust that you will enjoy using the units and the software.

Professional Development Unit for Geometry Expressions Using a Constraint Approach in Teaching Mathematics

Introduction

Geometry Expressions, developed by Saltire Software, is a computer application that, unlike other interactive geometry systems, can automatically generate algebraic expressions from geometric figures. A simple algebra system is embedded directly into Geometry Expressions, allowing the software to generate and simplify algebraic expressions for lengths, coordinates, areas, and equations for loci. An interactive symbolic geometry system affords a remarkable opportunity to blur the artificial line between algebra and geometry.

A constraint-based system is intrinsically different from construction-based. This is the most important thing to keep in mind while using Geometry Expressions. It has the capability of doing all of the constructions that a conventional geometry system can do, but at the same time it is actively making connections between the geometric objects you create. This is a powerful and useful function, as long as you know what is taking place! This is why we strongly recommend you work through this training package.

Structure of the training.

The training is organized into five short lessons that cover the basic capabilities of Geometry Expressions. It is not intended to be exhaustive. A firm grounding in the basics will allow you to understand the more complex behaviors of the software later, and to learn them as you need them.

Lesson 1: Using a Constraint Approach in Teaching Geometry

> What are constraints? How does a constraint-based or "symbolic" geometry system differ from other geometry software? How can my students take advantage of the differences, and what must they watch out for?

Lesson 2: Drawing Geometric Figures with Geometry Expressions

> How will my students draw figures in Geometry Expressions? How can they ensure their accuracy?

Lesson 3: Real Outputs and Symbolic Outputs

> What kinds of measurements does Geometry Expressions generate? In what sense is Geometry Expressions "symbolic?" How can my students use Geometry Expressions to bridge the gap between Algebra and Geometry? What links are there between Geometry Expressions and computer algebra systems?

Lesson 4: Variation and Animation

How can my students vary constraints to see the effect? How can my students use animation to illustrate a mathematical concept?

Lesson 5: Functions and Loci

How does Geometry Expressions represent functions? How can students define loci with Geometry Expressions? What advantages does Geometry Expressions have over graphing calculators?

We recommend that you work through the units either under the direction of an experienced user, or with partners in a self-directed way. We also recommend that you include time to work through some of the lessons available on www.geometryexpressions.com, or to create some lessons of your own.

Guide to the Screen

Here is a map of the Geometry Expressions Window. You'll learn its functions in context as you move through the training lessons.

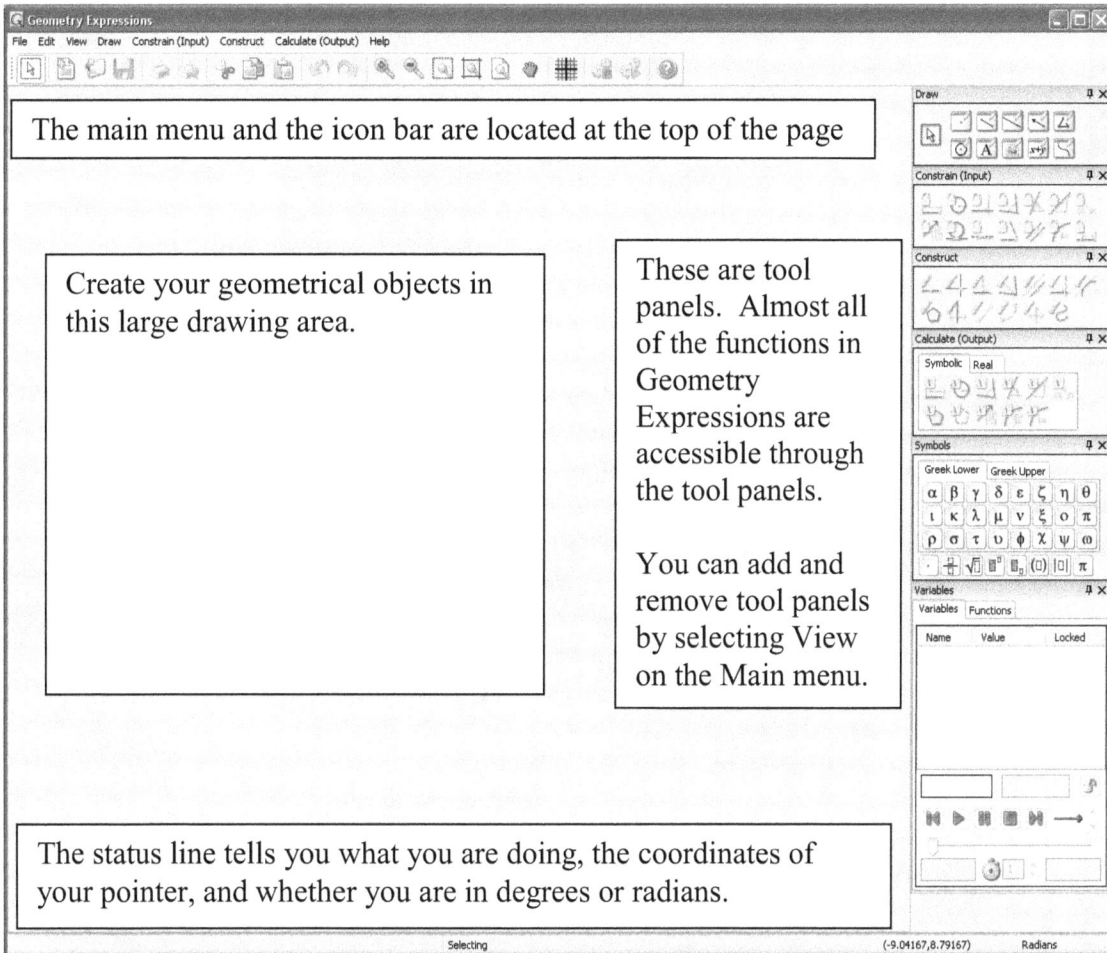

When you are asked to click on an icon in a tool panel, the name of the panel and the icon are written in bold, and a picture of the icon will often be included. For example:

Click on **constrain radius** .

If you "hover" the pointer over an icon, the name of the icon is displayed.

Menu selections are often written with slashes, in this format:

Edit/Preferences/Geometry/Line Color

Or as step-by-step instructions.

>Select Edit from the menu bar.

>Choose Preferences.

>Click on Geometry and change the Line Color to blue.

Lesson 1: Using the Constraint Approach

The intent of each Professional Development lesson is to explore how Geometry Expressions can be used to enhance classroom learning. Each unit addresses a particular feature of Geometry Expressions that may be different from other dynamic geometry tools that you have used. This lesson explores the concept of a **constraint system.**

A constraint system allows you to define geometrical objects in terms of other geometrical objects. Some of the constraints that you may impose on a geometrical object are

> A fixed length
>
> A fixed angle measure (including perpendicularity)
>
> A fixed direction or slope
>
> A fixed position
>
> Tangency
>
> Incidence
>
> Algebraic equation

These constraints are fixed in that they define other geometrical objects. You, the software user, can easily change them. The result is that the other geometrical objects are changed as well.

Keep in mind that mathematically, a good definition only includes characteristics that are **necessary and sufficient.** We generally have little trouble violating the "necessary" part of a definition – often we are still in the process of defining, or we wish to leave some parts undefined. I'm sure you've had this exchange in your classroom.

> "Draw a triangle."
>
> "What kind of triangle?"
>
> "It doesn't matter – any triangle will work."

The triangle created would be **underconstrained**.

Conversely, you could **overconstrain** the triangle, violating the "sufficient" part of the definition:

> "Draw an equilateral triangle with 60° angles."
>
> "Don't they have to be 60°?"
>
> "Excellent! Remember, class, that the angles of an equilateral triangle are all 60°."

Geometry Expressions will not allow you to overconstrain a geometrical object. It will remind you that your definition is already sufficient, and that any more constraints may or may not lead to a contradiction.

This lesson on Triangle Congruency Theorems is intended to exploit Geometry Expressions' detection of redundant constraints.

Nailing Down Triangles

The definition of a triangle is easy: it's a closed figure on a plane that is formed by three line segments. But how much information do you need to define a *particular triangle* – that is, one of a particular size and shape?

Open Geometry Expressions and get started!

First, make sure Geometry Expressions is in angle mode.

> Select Edit from the menu bar.
>
> Choose Preferences.
>
> Click on the Math icon on the right side of the dialog box.
>
> Toggle Angle Mode to Degrees.

1. **Draw** a triangle with the **segment tool** ⬚ or the **polygon tool** ⬚ . Don't forget to click on the **select tool** ⬚ when you are finished.

    Click and drag on parts of the triangle. You can see that you've created a triangle, but no triangle in particular. See if you can create the triangle in Diagram 1.

    Define the length of a side by selecting the side and then using the **constrain length** ⬚ tool.

    Define the measure of an angle by selecting both sides of the angle and then using the **constrain angle** ⬚ tool.

    You'll be done before you know it!

    Did you finish drawing the triangle? How far did you get?

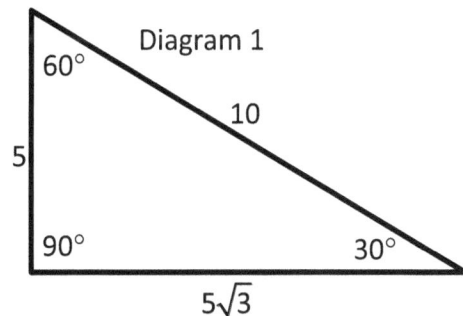

Diagram 1

(Diagram labels: 60°, 10, 5, 90°, 30°, $5\sqrt{3}$)

2. The reason that you got stuck before you finished is that Geometry Expressions won't let you tell it anything that it can figure out for itself. At first, this can be frustrating, but in fact it can be very powerful.

Look at the triangle in Diagram 2. What is the measure of the third angle?

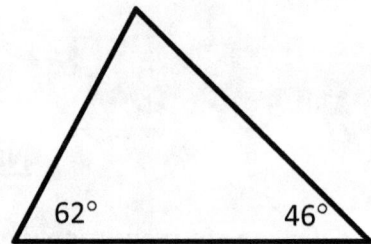

Draw the triangle in Geometry Expressions. Can you constrain all three angles? What happens if you try?

Diagram 2

Geometry Expressions won't let you constrain the third angle, because it knows it can figure it out. It lets you know with dialog box in Diagram 3. If you want to find out what the measure of the third angle is, choose "Calculate the angle from other constraints."

You can use Geometry Expressions to create a conjecture. Change the measures of your angles from 62° and 46° to x and y.

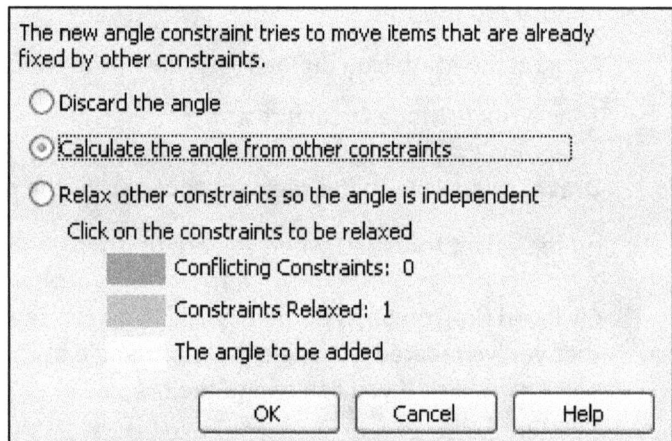

The new angle constraint tries to move items that are already fixed by other constraints.

○ Discard the angle

⦿ Calculate the angle from other constraints

○ Relax other constraints so the angle is independent

Click on the constraints to be relaxed

Conflicting Constraints: 0

Constraints Relaxed: 1

The angle to be added

| OK | Cancel | Help |

If the measures of two angles in a triangle are x and y, then the measure of the third angle

_____

Try dragging your triangle around now. You'll see that the angles change. That's because dragging is like typing in different numbers for x or y. Select x in the Variable Tool Panel, and click on the lock icon. That will keep the variable from changing when you drag. Do the same for y.

Drag the triangle around. Does your triangle keep its size and shape?

3. So far we've learned that you can only constrain two out of the three angles in a triangle, but that's not enough to nail down its size and shape.

Constrain sides until you get the dialog boxing complaining about your constrain. How many sides did you constrain *before* you got the dialog box?

Which side or sides did you constrain – between the angles, or not between the angles?

Compare your results with those around you. Sum up your findings here:

4. Start over with a new triangle. This time we're going to constrain sides instead of angles.

   Start constraining the sides. You can use symbols or numbers, but if you violate the triangle inequality your triangle will disappear. At each step, lock the variable and see if the triangle stays the same shape and size.

   Compare your results with those around you, and sum up your findings:

   Can you constrain an angle once all your sides are constrained? If you get the dialog box, click on "Calculate the angle from other constraints." Is it possible to find the measure of an angle in a triangle if you only know the lengths of the sides?

5. Continue your investigation with different combinations of angles and sides that you haven't yet tried.

   How many total constraints are necessary to define a triangle of particular shape and size?

   Are there any combinations that are exceptions to the rule?

6. One of the combinations of sides and angles does require one more bit of information to be complete: if you have two sides constrained, and the angle that is not between the sides is constrained.

Draw the triangle in Diagram 4 using Geometry Expressions. Constrain the sides and angle shown.

Make sure this angle <u>looks</u> obtuse.

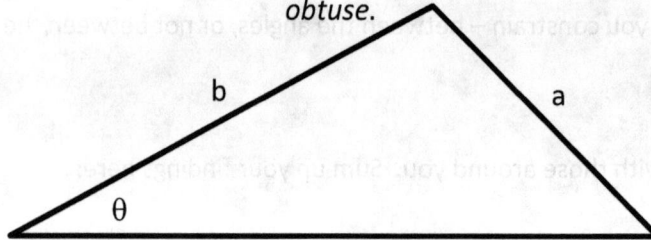

Diagram 4

Now, add a line segment as shown in Diagram 5.

How many triangles are there with angle θ and sides of length a and b?

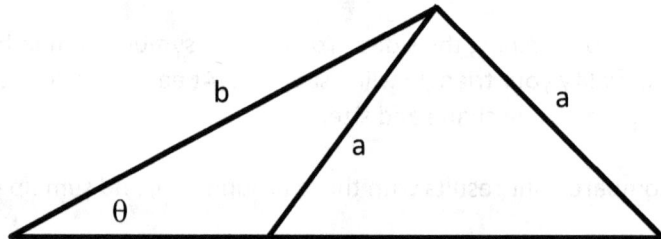

Diagram 5

Geometry Expressions uses your drawing to decide which triangle you intend.

7. Summary

You need _____ of the measures of a triangle defined before you establish its size and shape.

_____ does not work, since you can always find one angle from the other two.

_____ only works if you know

_____.

Extension:

Use Geometry Expressions to discover which combinations of sides and angles can be used to determine a quadrilateral's size and shape.

2009 Saltire Software Incorporated

Lesson 2: Drawing Geometric Figures

The primary purpose of this Professional Development lesson is to learn how to use Geometry Expressions to build geometric figures. This is, for the most part, a simple thing to do. It is more a matter of getting to know the software, and of remembering that Geometry Expressions is building a construction sequence. In other words, as you create geometric objects and constrain them, Geometry Expressions is in the background, building a list of instructions for creating your object.

Try each exercise intuitively. Then, read the discussion that follows, and try the different alternate solutions. It is in the trying of alternate solutions that you will learn how to use the software.

1. Draw a circle and a line segment that is its diameter.
   How did **you** solve this?

There are many ways to draw a circle and its diameter. It all depends on how you think about circles and diameters. We'll assume that you started with the circle, though there are certainly exceptions.

To draw a circle, click on the **draw circle** icon .

Click where you wish the center to be.

Click where you wish the edge of the circle to be.

Click the **select** icon , or you will be drawing nothing but circles!

a. The diameter of a circle is a chord that passes through the center of the circle.

First draw a chord:

Click on the **draw line segment** icon .

Click on the circle to make one endpoint.

Click another point on the circle for the other endpoint.

Select the chord and the center of the circle while holding down the shift key.

Click on **constrain incident** , which places the center on the chord.

b. Another way to draw a chord that passes through the center is to draw an infinite line through the center, and then find its points of intersection with the circle.

Click on the **draw infinite line** icon .

Click on the center of the circle (it will "highlight" when your pointer is on the center.)

Select the line and the circle while holding the shift key.

Click on the **construct intersection** icon .

c. The diameter of a circle is two radii that form an angle of 180° or π radians.

Use the **draw line segment** tool to draw to radii.

Hold shift and select both segments.

Click on the **constrain angle** icon .

There is an indicator in the lower right corner of the window that will tell you if you are in degree mode or radian mode. Type either 180 or click the π icon in the symbols tool panel.

d. The diameter of a circle is a chord with length twice the radius.

Select the circle and click on **constrain radius** . Leave the length of the radius as r.

Draw a chord.

Select the chord and click on **constrain distance** . Type 2*r.

Geometry Expressions allows variables to have more than one letter, so you must type an asterisk (*) if you wish to multiply.

e. Some will recall that the hypotenuse of a right triangle inscribed in a circle is the diameter of the circle.

Click on the **draw polygon** icon .

Click on three points on the circle to make an inscribed triangle. Click on the first of the three points again to close the triangle.

Hold the shift key and click on two of the sides of the triangle.

Click on the **constrain perpendicular** icon .

f. Ok, this one might be a little "out there," but remember we're trying to learn the features of Geometry Expressions. If two tangents to a circle are parallel, then the segment connecting the points of tangency is a diameter of the circle.

Select the circle and click on **construct tangent** .

Repeat, to construct another tangent.

Select both tangents by holding down the shift key and clicking on them.

Click on the **constrain parallel** icon.

Connect the two points of tangency with a line segment, and you have a diameter.

g. This one is probably cheating . . . .

Draw a line segment.

Construct its midpoint by selecting the line segment and clicking on the **construct midpoint** icon.

Draw the circle with center at the midpoint and circle on an endpoint.

2. Draw the incircle of a triangle. An incircle is a circle inside a triangle, tangent to each of its sides.

How did **you** solve this?

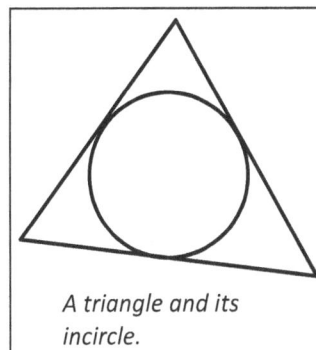

A triangle and its incircle.

Many students find the *construction* of an incircle a little tricky, especially since they forget to construct the perpendicular to the side from the incenter to get the radius.

a. The easy way to draw the incircle is to constrain the sides of the triangle to be tangent to the circle.

Draw a triangle with the segment tool.

Draw a circle inside the triangle.

Shift, and select a side and the circle.

Click on the **constrain tangent** icon .

Repeat for the other two sides.

b. The standard technique for constructing an incircle is to bisect two of the angles (locating the center), and then construct the perpendicular to one of the sides (finding the radius).

Draw a triangle with the segment tool.

Shift, and select two of the sides

Click on the **construct angle bisector** icon .

Repeat to bisect a second angle.

Shift and select both angle bisectors.

Click on the **construct intersection** icon.

Shift and select the point of intersection (the incenter) and one of the triangle sides.

Click on the **construct perpendicular** icon .

Construct the point of intersection between the perpendicular and the triangle side.

Draw a circle with center at the incenter and radius at the point you just constructed.

One of the advantages of Geometry Expressions is that you can create a figure based on its description and then explore its characteristics or you the other way around. It's your choice.

3. Draw a regular hexagon and find a formula for its area.

It would be wise to change the mode to degrees at this time.

Select **Edit** from the menu bar and choose **Preferences**.

Click on the **Math** icon on the left side of the dialog box

Choose **Degrees** for the **Angle Mode**.

How did **you** solve this?

This one can be challenging if you don't think it through and remember how the constraint system works. There are too many interdependencies for a "brute force" approach.

a. Start by constraining all six angles to a measure of θ . . . .

    Draw a hexagon with the line segment tool.
    Select two sides and click on **constrain angle**. Click on the θ icon in the symbol tool panel.

    This doesn't work because by the time you get to the sixth angle, Geometry Expressions has enough information to calculate it. The sixth angle is $720 - 5\theta$. You can "read the mind" of Geometry Expressions by clicking on the "Calculate the angle from other constraints" option in the error dialog box.

b. Ok then, start by constraining all six sides to be congruent . . . .

    Draw a hexagon.

    Select two sides and click on the **constrain congruent** icon ⬩ .

    Select one of the congruent sides and another side, and constrain those sides to be congruent.

    Repeat until all six sides are congruent.

    Unfortunately, we won't overcome the problem with the angle constraints. As a matter of fact, you may run into a collision after your fourth angle. If you click "Calculate the angle . . ." you'll see that it can be calculated from the sides and angles you've already constrained. It takes a while, and it's very complex, but it can be done.

c. Constrain *five* angles to be each 120° (the fifth will be calculated as 120°). Constrain *four* sides to be congruent. The other two sides can be calculated unambiguously through isosceles triangles.

    See previous step-by-step instructions, if necessary.

d. Inscribe the hexagon in a circle and constrain the sides to have the same length as the radius of the circle. This is the basis for a familiar compass-straightedge construction of a hexagon.

    Draw a circle.

    Draw a hexagon with vertices on the circle.

    Select the circle. Click the **constrain radius** icon ⬩ and type *r*.

    Select a side of the hexagon and constrain its length to *r*

    Repeat for the rest of the sides.

4. In a different vein, try drawing a line with a slope equal to 1, containing the point (3, 7). Click on the **toggle grid and axes** icon at the top of the page ▦ to see the coordinate axes.

How did you solve this?

a. You can constrain the coordinates of a point with **constrain coordinates**, and you can constrain the slope of a line with **constrain slope** 🖊 .

Draw an infinite line and place a point on the line.

Select the point and **constrain coordinates** 🖐 to (3, 7).

Select the line and **constrain slope** 🖐 to 1.

b. If you know the angle the line makes with the *y* axis, you can use **constrain direction** instead of constrain slope.

Delete your slope constraint from the previous diagram. If you try to constrain slope and direction, your line will be over-constrained.

Select the line and **constrain direction** 🖊 to 45 (are you still in degrees?)

c. You can plot the equation of the line $y = x + 4$.

Click on **draw function** 🗊 .

Select **Cartesian.**

Type in $x + 4$ at the **Y=** prompt

You should be ready to draw something more complex. The "First Ajima-Malfatti Point" is a point of concurrency in a triangle (Kimberling and MacDonald 1990, Kimberling 1994). Draw three circles inside of the triangle such that each circle is tangent to two sides and to the other two circles. Draw segments from each vertex to the point of tangency of the two remote circles. The three segments are concurrent at the "First Ajima-Malfatti Point."

The First Ajima-Malfatti Point

There are some tools that have not been addressed in this unit. Most are similar in nature to those you have already used. Some, like the function tool, the locus tool, and the transformation tools will be addressed in more detail in future lessons.

Lesson 3: Symbolic Outputs and Real Outputs

There are two methods of completing calculations in Geometry Expressions: Symbolic and Real. If you examine the Calculate Tool Panel, you'll see that there is tab for each mode. Click on the Symbolic tab and then compare with the Real tab, and you'll notice very little difference – all calculations can be done either symbolically or as real number approximations. If you closely examine the icons, you'll see that the symbolic versions have an x: , while the real versions have an 88: .

The first problem deals in real number approximations, so select that tab in the calculate tool panel.

1. A farmer wishes to use 100 feet of fencing to enclose a rectangular area. Being clever, he plans to use the fence for three sides of the area, and an existing wall of sufficient length for the fourth wall. What is the maximum area of the, uh, area?

   Use your newly developed Geometry Expressions talents to draw the rectangular area.

   Remember that the software doesn't solve your problem for you. "How do I get it to figure out an expression for the length of each side?" is probably the wrong question – that part is your job.

   Be careful not to overconstrain the rectangle. What is a good definition of a rectangle?

   It is very likely that your rectangle no longer fits nicely in the window.

   Click the **scale down** icon at the top of the screen until you see the entire rectangle.

   Modify the domain of your variable with the variable tool panel, shown at the right. The example uses x for the length of a side, and [0, 100] for a reasonable domain on x.

The boxes at the bottom of the panel control the domain of x. The box next to the stopwatch controls animation

Select all four sides of the rectangle, and click on **construct polygon** .

Select the interior of the rectangle and click on **calculate real area** .

What is the maximum area? Change the value of your variable with the slider bar in the variable tool panel. When you get close, narrow the domain of your variable and see if you can get closer.

To get more precision:

 Select the area calculation.

 Right click on the calculation.

 Choose Properties from the pop-up menu.

 Change the Decimal Digits.

 Delete the area calculation, select the interior of the rectangle, and click on **calculate symbolic area** . You'll see one of the following, or something that is algebraically equivalent:

$$z_0 \Rightarrow |100 \cdot x - x^2|$$

 If you see this, you have "use assumptions" set to false. Geometry Expressions isn't taking any chances, and it uses the absolute value signs to make sure the area is positive.

$$z_0 \Rightarrow \frac{100 \cdot x - x^2}{|x < 100}$$

 If you see this, you have "use assumptions" set to true. Geometry Expressions is assuming that since $100 - x$ represents a length, that $x < 100$.

To change the setting for "use assumptions" for a particular calculation:

 Select the calculation.

 Right-click the calculation.

 Choose Properties from the pop-up menu.

 Toggle the value for Use Assumptions.

You can also set the default value for "use assumptions" through the Preferences dialog box.

Click Edit on the menu bar.

Choose Preferences.

Click on the Math icon at right.

Toggle the value for Use Assumptions.

Back to our farmer problem – the symbolic area expression can be used by students to determine a maximum (it's the vertex of the parabola). These expressions can also be exported into a variety of computer algebra systems if desired.

2. Which is greater: the ratio of the area of a circle to the area of its circumscribing square, or the ratio of the area of a square to the area of its circumscribing circle?

Complete your drawings and let Geometry Expressions calculate the areas.

If you did completely constrain your figures, you'll see some variables that you did not create in your expressions. These are *system added constraints.* Geometry expressions will create temporary constraints to finish calculations if it needs to. If you add more constraints of your own, expressions will be recalculated to take advantage of them.

You can use the Draw Expression tool to create new expressions. For example, you can create the ratio of two areas.

Click on **draw expression** icon ![x+y].

Type in the ratio of your two areas expressions, enclosing the subscripts in square brackets. For example, if the area of my square was z_0, and the area of my circle was z_1, for the ratio of square area to circle area I would type in z[0]/z[1].

If you did the exercise with *real* calculations, delete all of you area calculations and expressions, and re-build then as *symbolic* calculations.

3. Find the symbolic area of a triangle.

It is best to be in radian mode for this exercise.

First, draw a triangle without constraining any of its sides.

The symbolic area will be in terms of *system added constraints.*

Click on the area expression and the system added constraints are shown on the diagram.

Add one constraint to the triangle, either an angle measure or a side length.

The area expression will change to include your constraint. It may redefine the system added constraints as well.

Add a second constraint, and then a third constraint. You may catch a glimpse of the Law of Cosines, or Herons' formula.

4. If the lengths of the sides of a triangle are *a, b,* and *c,* what is the radius of the incircle in terms of *a, b,* and *c*?

Lesson 4: Variation and Animation

One of the more useful features of GX is the ability to manipulate specific constraints and variables within a drawing. We will introduce those features here, and then use them extensively in working with functions and loci.

1. A simple but illustrative example.

    Draw a circle, and constrain its center with **constrain coordinates** .
    Use the default coordinates (x_0, y_0).

    Constrain radius to r.

    Click and drag the points on the diagram, and see how different combinations of constraints change values depending on what you drag.

    What if you want to modify only the x coordinate, or only the radius? You have a couple of options.

    a. Your first option is to lock variables that you don't want to change.

        Suppose you want to relocate the center of the circle, but don't want the radius to change.

        Look in the variables tool panel, and you'll see all three constraints listed.

        Highlight r and then click on the padlock icon. This locks r, so it won't change as you drag elements of your diagram. Now drag the center point of your circle.

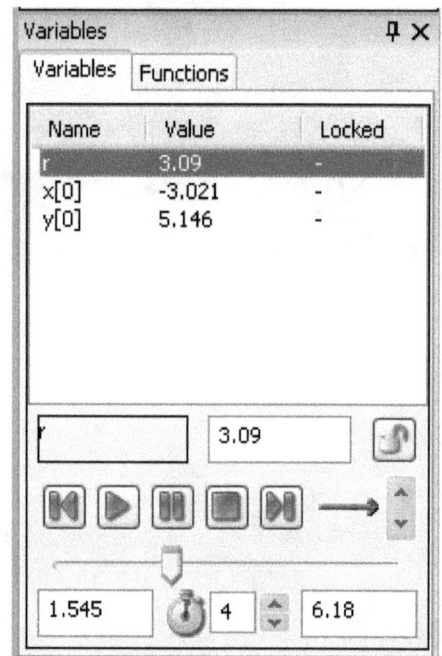

    b. Your second option changes one specific constraint at a time.

        Unlock r by clicking on the padlock icon again. Highlight x[0], then look at the bottom of the window. Drag the slider bar, and see what effect it has on the diagram. The values in the bottom left and right give the minimum and maximum values for the slider bar.

        Modify them and test the effect of dragging the slider bar.

        Do the same thing in turn with y[0] and then with r.

        You can manipulate the value of any constraint in this way.

c.   The third option is animation.

You can use animation to move through values of a variable. This automates what you just did with the slider bar. Choose a variable, enter a minimum and maximum value, and click on the play button. You'll notice that the diagram goes through all the specified values for the chosen constraint smoothly.

2.   A higher-level example.

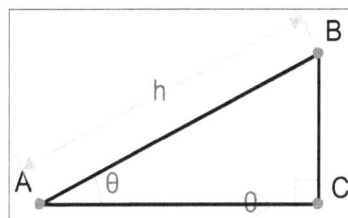

Draw a right triangle.

Constrain angle C constrained as the right angle.

Constrain angle A to measure θ,

Constrain AC to slope 0

Constrain AB (the hypotenuse) to length h.

Make sure point B is above $\overline{AC}$, and your computer is set to degree mode.

Lock the hypotenuse length, and drag point B in the diagram to see the variations of your triangle. Dragging allows you to very quickly show students multiple examples of a figure with given constraints.

Practice using the slider bar and the animation features to a similar effect. Notice that the default for θ is 0° through 360°. Change the upper bound for animation to 90° to demonstrate triangle trigonometry, or keep it at 360° to demonstrate circular functions.

You may have noticed the blue arrow to the right of the animation controls. This indicates that the animation will run from the minimum value to the maximum value, and then the animation will stop.

Click the up and down arrows to the right of the blue arrow to investigate the other animation modes.

You can change the speed of the animation by changing value next to the stopwatch icon. The value represents the number of seconds that it will take to complete one iteration of the animation. The maximum value is 60 seconds.

Lesson 5: Functions and Loci

One of the many useful tools in Geometry Expressions allows students and teachers to create function graphs, and then change them dynamically. While some of this functionality can be found in graphing calculators, the dynamic changing and ease of connecting with other representations sets Geometry Expressions apart.

1. We start by graphing a family of basic quadratic functions $y = ax^2$.

   Use the Draw/Function tool, and type the equation in under the Cartesian option. Remember that you must us * for all multiplication, and ^ for exponents. Make sure your axes are turned on.

   Use the tools in the Variables window to modify and animate a. You can very quickly see the effect of changing values of a have on the graph of the function. Modify the range of values for a to include negative numbers, and you can see the smooth progression of a single pattern.

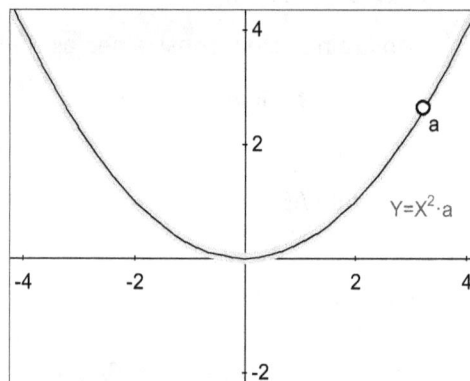

   Finally, highlight the curve, then click and drag it. A "handle" will appear with the label a. You can vary values of a by dragging that handle.

   Edit the equation by double clicking on it, and make it $y = ax^2 + b$.

   Use the variables window to vary values for b, and observe the effect.
   Click and drag on the curve twice, and get handles for both a and b. Students can very quickly and intuitively see the effects of vertical translations and dilations this way.

2. Standard geometric transformations can be done to functions as well as any other object in Geometry Expressions.

   Open a new file, and graph $y = 0.5x^2$.

   Use **Draw vector** to create an arbitrary vector.

   Note that you can drag either endpoint to change magnitude and/or direction, but if you highlight the whole vector, dragging it to a different position leaves it unchanged.

   Constrain Coefficients to give values of $\begin{pmatrix} u_0 \\ v_0 \end{pmatrix}$ to the vector.

Select the graph of the function, and click on **Construct Translation** ✏️ . Then select the vector.

Drag the endpoints of the vector around, and watch the translation change dynamically.

To more clearly discern between the original parabola and its image, you can change the display properties of the image.
Select the image parabola, and then right-clock on the image parabola.
Select Properties/Line Color, and choose whatever appeals to you.

You may wish to see how other quadratic functions are related by translation.
You could create a new function with the equation
$y = 0.5(x - 4)^2 - 2$. Then drag the vector endpoints until the image coincides with the new function.

Repeat the process, but make the second function
$y = 0.5x^2 + 2x + 5$.

Finally, lock your vector coefficients in the variables window, and constrain the tail of your vector to be incident with $y = 0.5x^2$. Drag the vector around, and you can see a direct, point-by-point correspondence between the two functions. We will return to parabolas in the next section.

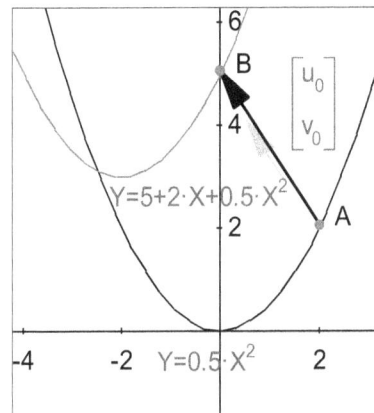

This type of activity can serve as part of a bridge between standard and vertex forms of quadratic equations, as well as demonstrating the effects of translations.

3. Besides Cartesian equations, Geometry Expressions will also graph polar and parametric functions.

    Open a new file.

    Draw a **function**, but choose **parametric** for the type.

    Enter the equations $X = a * \cos(T)$ and $Y = a * \sin(T)$.

    You should get the circle as expected, with a variable radius of length a.

    Select the circle, and click on **Calculate Real Implicit Equation** 📊 to produce the Cartesian equation for the circle.

    Calculate Symbolic Implicit Equation, will give you the general form for the equation of a circle, in terms of the constraints in the diagram, namely the radius a.

Now edit your function to $X = a^* \cos (T)$ and $Y = b^* \sin (T)$.

As you vary a and b values, you get an ellipse. Students can see dynamically how those constraints impact the appearance of the graph, and how a circle could be considered a special case of an ellipse.

Now think parametrically in terms of projectile motion. Recall that the parametric equation for a projectile shot at an angle of 55°and a speed of 200 feet/second is given by

$$\begin{pmatrix} x = 200\cos(55)T \\ y = 200\sin(55)T - 16T^2 \end{pmatrix}$$

Create a new Geometry Expressions file, and check to see that you are in degree mode.

Draw the parametric **function.**

Set the **Start** and **End** values for T to 0 and 15, respectively.

You will need to scale out ([icon] on the top toolbar) to see the graph. Scaling up and down, true to its name, adjusts the scale on your axes – the labels stay the same size and font. If you want to truly zoom in or out – magnifying or shrinking the whole picture proportionally, use View/Zoom In or View/Zoom Out. Try both ways to see the difference. Zooming may be especially useful if you use a classroom projector and want to make a diagram easy to read in the back of the room.

One advantage we have is that we are not limited to a static picture. To draw a point that is snapped to the graph:

Draw a **point** on the graph.
Select the point and the curve together, and select the **Constrain Point**

Proportional [icon] tool.

This feature does slightly different things, depending on context. In this case, the point will be on the parametric function curve at a position corresponding to time t.

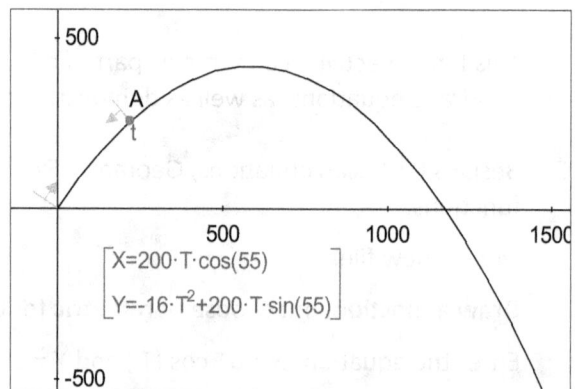

We can now animate t in the variables window and/or drag the slider bar, and watch the path of the projectile at specific times. More sophisticated uses of this feature can empower student to determine if two moving objects will collide, or how to modify their trajectories to ensure that they do.

4. The ability to create a locus is one of the more powerful tools in GX.

A locus is a set of points that meet given criteria. We'll start by looking at parabolas again – this time from a geometric definition.

A parabola is the set of all points equidistant from a fixed point (the focus) and a fixed line (the directrix). There are many ways to construct this. One of the simpler options follows:

Draw a **point** A and an **infinite line**.

Constrain the **coordinates** of the point to the default: (x_0, y_0).

Constrain the line's **implicit equation** to be $y = 0$.

Draw a second **point**, B.

Draw line segment $\overline{AB}$ and a line segment from B to any point C on the directrix.

Select $\overline{BC}$ and the directrix, and click **constrain perpendicular** .

Select both $\overline{AB}$ and $\overline{BC}$. **Constrain** the segments to be **congruent** .

Drag C back and forth, and point B will follow the curve of a parabola.

To trace this path, we need a parameter that will control the position of point C.

Point C is already partially constrained, since we placed it on the directrix. That's why we can't use constrain coordinates on point C.

Instead, select C and the line, and use **Constrain Point Proportional....** ,
The parameter (default t) gives the x-coordinate of point C as it moves along the line.

Adjust the range of values for t, and animate to see the curve followed again.

Now select point B, and click on **Construct Locus** . Select parameter t, and give an appropriate start and end value.

Geometry Expressions now traces the path of point B.

$$Z_0 \Rightarrow -2 - 0.125 \cdot X^2 + Y = 0$$

$Y = 0$

$A\left[x_0, y_0\right]$

You can change your parabola by moving the focus, either by dragging or through the variable tool panel. Try changing the position of the line by editing its equation. The locus adjusts to each new change.

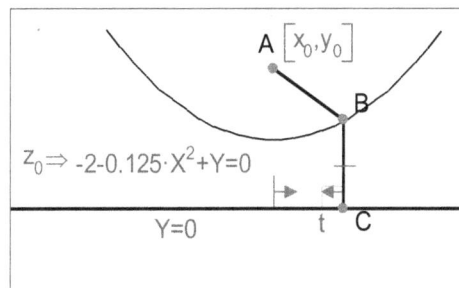

Select the locus, and **calculate** its **real implicit equation**. Your students can use this equation to see how the functions changes as the diagram changes. With some guidance of suggested focus and directrix positions, students may be able to make generalizations on their own. For example, set the coordinates in the variables window to (0, 0.25), and the equation of the directrix to y = -1*y[0]. This puts the focus and directrix equal distances from the x-axis. What happens as you increase y_0 with the slider bar?

Similarly, students can plot a generic quadratic function with the function tool, and then change its parameters to match the locus curve.

An ellipse can be similarly constructed from its locus definition, the set of all points whose distances from two foci equal a constant. Try to construct it independently before reading on.

Draw two points A and B anywhere on a blank diagram, and constrain their coordinates to the defaults. Draw a third point C, and constrain its distance from one of the foci to be *a*, and its distance from the other to be some constant minus *a*. The constant you choose must be greater than the distance between the points. Construct the locus of point C, using *a* as the parameter. The difficulty is that we can't effectively define a parametric variable that will give us the whole rotation, so we get half an ellipse. The simple solution is to draw a line through A and B, and then reflect the locus through it. To do the reflection, highlight the locus, select the Construct/Reflection tool, and then select the line. Notice that the foci can be dragged around – with the cursor or in the variables window – and the rest of the diagram follows and adjusts.

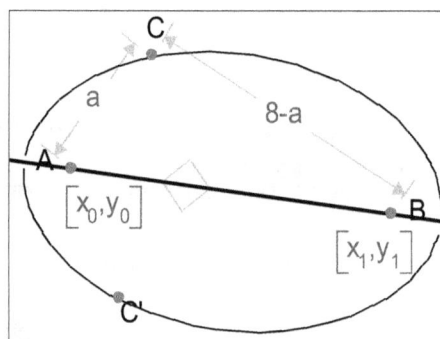

5. Connections and Simulations

Some powerful learning tools can be created by combining the techniques you've learned so far. For example, one common difficulty students can have in trigonometry is transitioning from the idea of coordinates of a point on a unit circle to a function graph. This idea can be made clear with a dynamic geometry tool like Geometry Expressions.

Open a new Geometry Expressions file, and make sure you are in radian mode.

Draw a circle centered at the origin and **constrain** its **radius** to one.
Constrain point B **proportional to** the curve, and set the parametric variable to θ. You can find θ in the symbols tool panel.

This puts B on the unit circle, at a rotation of θ from the positive x axis. It may help some student to visualize better if you draw line segment $\overline{AB}$ and constrain its angle to the positive X axis. Draw a third point at a blank spot, and constrain its coordinates to $(\theta, \sin(\theta))$.

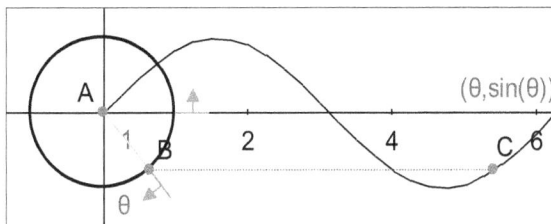

Drag point B or C around, and/or vary θ in the variables tool panel

Drawing a line segment between B and C sometimes helps with the visualization.

Construct the **locus** of point C, and you can see the sine curve.

Through animation, a person can see dynamically how the function curve is created from the height of a point on the unit circle after a rotation of θ radians.

Students at a higher level, can do a variety of transformations to the sine function to model various situations, primarily circular and harmonic motion. Consider the problem of the position of a rider starting at the bottom of a Ferris wheel centered 40 feet above the ground with a radius of 35 feet, which makes one revolution every 32 seconds.

First make a "physical" model of the ride.
Draw a circle, and constrain its center (A) and radius appropriately. To make the point B move as desired is a bit trickier. Draw a line segment from A to B, then constrain the direction of the segment. This constraint must take into account the period of 32 seconds, and the 8 second time-shift represented by the rider starting a quarter turn before the normal starting point of the sine function. The appropriate expression is $\frac{2\pi}{32}(t-8)$.

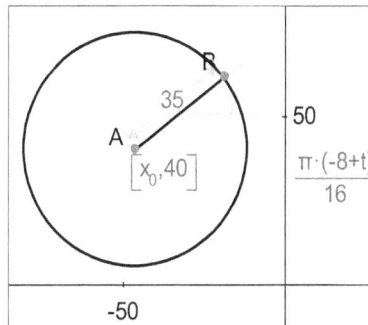

Animate point t for two cycles (64 seconds) and watch the ride simulation.

We transform this into a function graph in a manner similar to what you did for the unit circle.

Draw a point C and constrain its coordinates to $\left(t, 35\sin\left(\frac{2\pi}{32}(t-8)\right)+40\right)$.

Animate and then construct the locus, and animate again. Remember the different animation options if you want the ride to keep going. Students can see the function graph "drawn" for them simultaneously with the animation of the physical model.

Appendix: Troubleshooting

Problem: Objects disappear

> **Reason:** You can create objects that are impossible to draw. For example, if you constrain the sides of a triangle to be lengths a, b, and c, and later assign values $a = 5$, $b = 6$, and $c = 15$, only the side of length c will be drawn

> **Solution:** Ctrl-Z or the undo icon 🔄 will undo your last change. You can undo as many steps as you need. You can also use Ctrl-Y or 🔄 to reinstate a change

> Try using the slider bar in the variables window to modify constraints in your drawing.

> Scale down 🔍 your drawing to see of the object has moved off the screen

Problem: Angle measures don't appear as expected

> **Reason:** Most likely, you haven't set GX to the angle mode (degrees or radians) you want.

> **Solution:** Change the angle mode. Select **Edit** from the menu bar and choose **Preferences.** Click on the **Math** icon at the left of the window. Change **Angle mode.**

> **Reason:** If you are using the direction constraint, GX takes into account the order in which the line segment was drawn.

> **Solution:** Delete the line segment, and re-draw it in the other direction. Another solution is to add 180° or π radians to the direction. You may also decide to use "slope" instead of "direction" as your constraint.

Problem: Cannot type π when creating a function or setting the value of a variable.

> **Reason:** This is a temporary problem.

> **Solution:** You can use the π symbol when you are editing a function, but not while you are creating it. Create the function, and then double-click on it to edit it. If you wish to use π in the variable tool panel, you must use an approximation for now.

Problem: You create a locus, but do not see what you expect.

> **Reason:** You have set the wrong parametric variable, or the domain of the parametric variable is not set correctly.

> **Solution:** Double-click on the locus you see, and change the parametric variable and/or its start and end value. If you cannot see the locus at all, undo the creation of the locus and start again.

Reason: Other parts of the diagram are unconstrained. For example, your locus is defined in terms of a fixed line, but you haven't fixed the line.

Solution: Add the necessary constraints to the drawing.

Reason: Your locus definition includes multiple points for the same value of your parametric variable. For example, you only see half of a parabola or half of an ellipse.

Solution: You can reflect your locus across an axis to see the rest of the points.

Applying Geometry Expressions in the Algebra 2 and Pre-Calculus Classrooms

Unit 1: Parametric Functions

Primary Mathematical Goals

- Students will review the definition of a Cartesian function (y as a function of x).

- Students will be able to model real problems with parametric functions.

- Students will use animation to see how functions with the same "graph" can be different in terms of motion.

Overview

When students see a graph representing projectile motion, it is easy for them to confuse the graph with the actual path of the projectile. That's because the horizontal axis represents the passing of time rather than a horizontal displacement.

Parametric functions provide an easy mechanism for creating a graph that more accurately reflects the physical position of a projectile. Similarly, it portrays circular functions as circles, rather than sine curves.

Still, when looking at a static graph of a parametric function, the sense of movement is lost. Students wonder, "Where is *t* on the graph?" Computer technology provides the sense of movement, allowing students to connect *x, y,* and *t*.

The main goals of this unit are to review functions in general, to familiarize students with parametric functions and their applications, to see how motion is included in the representation of a parametric function, and to uncouple the idea of the vertical line test from the definition of functions in general.

Outline

Lesson 1: Functions: A Quick Review   *Time Required: 30 – 45 minutes*

- The definitions of function, control variable, and dependent variable are reviewed. Students use Geometry Expressions to see how the vertical line test relies on the correspondence between *x* and *y* on the graph.

Lesson 2: Dude, Where's My Football?   *Time Required: 60 – 75 minutes*

- A problem about the angle required to kick a field goal is used to introduce parametric functions. Geometry Expressions is used to show the motion of a point in two dimensions with respect to time. This naturally evolves into parametric functions. The idea that the parameter, *t*, only appears on the graph through the motion of the point is emphasized.

Lesson 3: Go Speed Racer   *Time required: 90 – 120 minutes*

- The idea of motion in the graph of a parametric function is further explored. Functions that have the same *x-y* graph but different parametric equations are examined. The idea that the vertical line test does not apply to parametric graphs is again stressed.

Lesson 4: Parametric Problems   *Time required: 90 – 120 minutes*

- Students are asked to apply the parametric function to problems involving projectile motion.

Lesson 1: A Quick Review of Functions

Learning Objectives

This is the first lesson in the unit on parametric functions. Parametric functions are not really very difficult – instead of the value of *y* depending on the value of *x*, both are dependent on a third variable, usually *t*. Confusion can occur when students try to use Cartesian approaches to solve problems with parametric functions. For example, a student might decide that a parametric graph does not represent a function, because it does not pass the vertical line test.

Before we contrast parametric functions with Cartesian functions, we must first review our understanding of Cartesian functions. That is the purpose of this lesson.

Math Objectives

- Review the concept of a function as a mapping from one variable to another.
- Review control variables and dependent variables.
- Review the vertical line test and its role in determining whether a graph is a function.

Technology Objectives

- Use Geometry Expressions to create and graph functions, and constrain points to functions.

Math Prerequisites

- Previous knowledge of the definition and properties of functions is helpful.

Technology Prerequisites

- None

Materials

- Computers with Geometry Expressions.

Overview for the Teacher

1. In part one, students walk through the creation of a parabola in Geometry Expressions, and then placing a point on the parabola.

   Diagram 1

   If students are able to move the point off of the curve, they created a point and then dragged it to the parabola. Thus, it is not part of the parabola. Help them follow the directions more carefully.

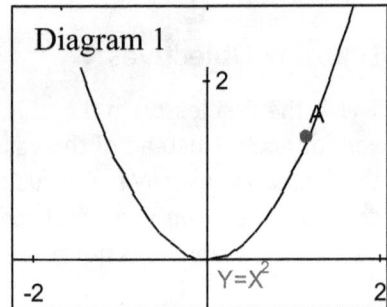

   Diagram 1 is representative of student work.

2. When students move the mouse pointer to the left and right, the point will move along the curve. When they move the pointer up and down, the point does not move very much. In fact, the amount it moves reflects the amount the pointer movement varies from the vertical. If student have trouble with this, have them trace the mouse pointer up the *y* axis.

   This is supposed to demonstrate that *x* is the control variable. This lesson uses the term "control variable" for *x*, because it fits nicely with the "controls" students use to manipulate the variable in the software. You may wish to connect this term with one of its synonyms: independent variable, input variable, or manipulated variable.

   Check to see that students are "controlling" the position of the point by changing the *x* coordinate.

   The results of **Calculate Symbolic Coordinates** will be (x, x^2). Diagram 2 shows typical student work.

   Diagram 2

   The second coordinate depends on *x*.
   y is the dependent variable.

3. Question 3 reinforces the vertical line test. Students drag a vertical line across the parabola. Then, they attach the line to their point, and animate the movement of the point and line together.

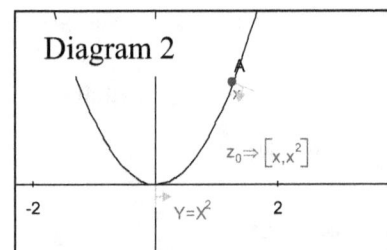

   You may wish to encourage students to zoom out to see more of the graph before forming a conclusion.

   The parabola does pass the vertical line test.

4. In part four, students are presented with several functions they may or may not be familiar with. Students apply the vertical line test to each function.

Note that sqrt(X) represents $\sqrt{x}$ and that 2^x represents 2^x.

	Coordinates of Point	Pass vertical line test?	Is it a function?
$y = x^2$	(x, x^2)	YES	YES
$y = \sin(x)$	$(x, \sin(x))$	YES	YES
$y = \sqrt{x}$	$(x, \sqrt{x})$	YES	YES
$y = \ln(x)$	$(x, \ln(x))$	YES	YES
$y = 2\^x$	$(x, 2^x)$	YES	YES

Note also that the software uses log for natural log, which is more common in computer and engineering fields.

5. Summary:

Functions in this lesson are sets of points with these properties:

x is the control variable.

y is the dependent variable.

A rule (or formula or equation) explains how x and y are matched up.

Every value for the control variable corresponds with only one value of the dependent variable.

The graph of the function passes the vertical line test.

Student Worksheets

Student worksheets follow.

A Quick Review of Functions

We are about to look at functions is a new way: in a form called "Parametric Functions."

But before we do that, we need to quickly review the old way!

Start by opening a new file with Geometry Expressions. Turn on the axis (look at the icon bar along the top) if it is not already on.

1.   Create a parabola.

Click on the **Draw Function** icon

Leave the type as *Cartesian*

Type x^2 in the Y= box

You should see the familiar parabolic graph of $y = x^2$

Now draw a point on the graph.

Click on the **Draw Point** icon

Move your cursor over the parabola. It will change color when you are directly over it.

Select the parabola.

Click on the **Draw Select** icon when you are finished drawing your point.

Sketch your parabola

2.   Who is in control?

The most important characteristic of a function is that one variable is in control – the control variable – while the other depends on the value of the control variable.

Click on the point and carefully drag your cursor to the left and right. What happens to the point?

Now carefully drag up and down. What happens now?

Which variable is in control, x (left and right) or y (up and down)?

Hold down the shift key and select the point, then select the parabola.

Click on **Constrain Point proportional along curve.**
Change the variable from t to x.

x is now listed in the Variable Tool Panel, though y is not. Select it, and change the animation control box to look like this:

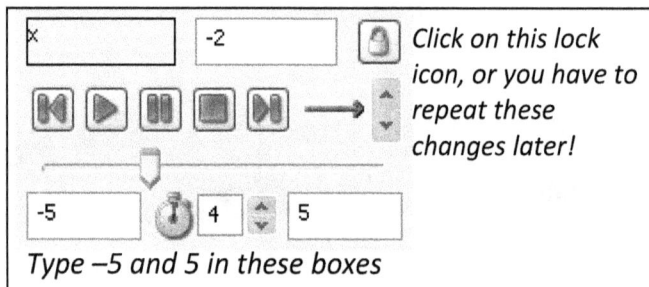

Click on this lock icon, or you have to repeat these changes later!

Type –5 and 5 in these boxes

Drag the slider bar, or click on the play button to see how x controls the position of the point.

Now, select the point.

Click on **Calculate Symbolic Coordinates.**

What are the results?

What does the second coordinate depend on?

Which variable is dependent?

3.   It is often said that a graph is a function if it passes the vertical line test. Does the parabola pass the vertical line test?

Click on **Draw Infinite Line** and draw a line in the window.

Click on the **Select Arrow** when you are finished drawing the line.

Select the line and Click on **Constrain Direction.**

Look in the lower right corner of the Geometry Expressions window. It will tell you if you are in degrees or radians.

        If you are in degrees, type 90.

        If you are in radians, click on the π icon in the symbols window, then   type /2.

Drag the line across the parabola. Does the line ever cross the parabola more than once?

Hold down shift, select the point, and select the line. Click on **Constrain Incident.**

This will place the point on the vertical line.

Select variable *x* from the Variables Tool Panel. Click Play.

The definition of a function is this:

> A function is a mapping between two sets such that each member of one set corresponds with only one member of the other set.

The vertical line test demonstrates that each value of *x* corresponds with only one value of *y*.

Does the parabola pass the vertical line test?

4.   Let's look at some other functions.

First make sure you are in Radian mode.

        Click Edit on the menu bar.

        Click on Preferences.

        Click on Math.

        Change Angle Mode to Radians.

Look at the Geometry Expressions window and find the equation of your parabola: $Y = X^2$

        Double click on it.

        Type sin(x)

What are the coordinates of the point now?

Select *x* in the variables window and press play.

Does y = sin(x) pass the vertical line test?

Repeat with sqrt(x), ln(x), and 2^x

(Hint: Make sure that your point is in quadrant I or IV (so that x > 0) before changing the function equation. Otherwise, it may disappear).

	Coordinates of Point	Pass vertical line test?	Is it a function?
y = x^2			
y = sin(x)			
y = √x			
y = ln(x)			
y = 2^x			

x^2, sin(x), $\sqrt{x}$, ln(x) and 2^x are all **rules** for finding the y coordinate. Every function must include a rule that explains how the control variable is used to calculate the dependent variable.

5. Summary:

Functions in this lesson are sets of points with these properties:

x is the _____ variable

y is the _____ variable

A _____ explains how *x* and *y* are matched up.

Every value for the _____ variable corresponds with only one value of the _____ variable.

The graph of the function passes the _____ test.

Lesson 2: Dude, Where's My Football?

Learning Objectives

Many students expect a "falling object" graph to look just like the path of the falling object, but that isn't usually the case. The graph simply shows the height of the object with respect to time. No information about the horizontal displacement is shown on the graph.

One of the virtues of a parametric function is that it gives a "true picture" of the position of the object over time. One of the drawbacks is that the parameter does not appear on the graph.

This drawback is addressed through computer geometry systems, but that's a topic for the next lesson. We must learn what parametric functions are first.

Math Objectives

- Define a "parametric function."
- Apply parametric functions in context.

Technology Objectives

- Create parametric functions with Geometry Expressions.
- Constrain points to the graphs of parametric functions to see how they move.

Math Prerequisites

- The concept of *functions.*
- Distance = rate multiplied by time.
- Functions for heights of falling objects.

Technology Prerequisites

- Concept of zooming and panning a display.

Materials

- Computers with Geometry Expressions.

Overview for the Teacher

The introductory problem is not to be solved straight away – its purpose is to create a need for parametric functions. The question "where is the ball" requires a two-part answer: how far has the ball moved horizontally, and how far has it moved vertically.

x stands for the horizontal displacement, and is controlled by t, the amount of time since the ball was kicked.

1. First, the horizontal displacement is addressed. Since nothing influences the horizontal movement of the ball except for air friction (which is disregarded) and impact with the ground (which creates the upper bound of the domain), the function for x is rate * time.
$X = 20t$.

Following the directions carefully and hitting play will result in point A traveling along the x axis at a uniform velocity.

Responses to "How does t show up on the graph" might be:

Nowhere – only x and y appear on the graph, or

Implicitly in the values for x and y, or

t appears in the movement of the point.

2. In part 2, the function for vertical displacement is explored.

If students are unfamiliar with falling bodies problems, you will need to step through this part carefully as a class.

The formula for y is $-16t^2 + 30t$

After directions are followed, point B will move up the y axis, slowing to a stop, and then moving down the y axis.

3. Part 3 combines the results of parts 1 and 2 to give a more realistic picture of the path of the football.

Point C will move in the parabolic arc that describes the path of the ball in two dimensions. Students may take one of two approaches to answering the question "where is the ball after 1.5 seconds?" Students may

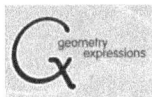

Move the slider or type in 1.5 for *t* in the Variable Tool Panel, and then read
approximate numbers from the graph or
Substitute 1.5 for *t* in each formula.

Again, *t* appears on the graph implicitly in *x* and *y* and in the movement of the point.

4. Part 4 introduces the parametric function notation, as implemented by Geometry
Expressions.

You will want to show your students the standard format:

$$f(t) = \begin{pmatrix} x = 20t \\ y = -16t^2 + 30t \end{pmatrix}$$

After completing the task, students will see the parabola itself, and the point will move
along it.

5. These problems will help reinforce the concepts in this unit.

    a. 468.8 feet in about 3.85 seconds.

    b. About 8.9 feet away from Jackie, and about 12.25 feet high.

    c. Yes. The new record will be 76.04 feet.

6. Summary:

A parametric function describes both x and y in terms of a third variable called the
parameter.

The parameter is the control variable for a parametric function.

On an *x-y* graph, the value of *t* appears only in the motion of the point through the curve.

Student Worksheets

Student worksheets follow.

Dude, Where's my Football?

Jerry is practicing kicking field goals. Jerry kicks the ball so that it has a horizontal velocity of 20 feet per second, and a vertical velocity of 30 feet per second. Where is the ball after 1.5 seconds? (Disregard air friction and other nominal influences!)

In the last lesson, we reviewed functions and found that for a function:

> *x* was in control

> *y* depended on the value of *x*

What does *x* stand for in the example above?

Is *x* in control, or is something controlling *x*?

Fill in the blank: The distance the ball has traveled horizontally depends on _____

_____

══

1. *x* as a function of *t*

    If you thought that *x* was controlled by the amount of time since the ball was kicked, then you were absolutely right! *x* is a function of *t*.

    First, let's find the rule for how *x* is calculated as a function of time.
    Remember that distance = (rate)(time), and that horizontal velocity is 20 feet per second.

    Complete the formula: *x* = _____

    Open a new file in Geometry Expressions.

    Turn on the axis.

    Scale down with the icon on the top icon bar 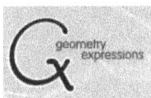 until the scale includes 50.

    Create point A.

Select the point and click on **Constrain Coordinates.**

> Type the expression that you wrote in the blank. Use * for multiply.

> Type a comma, then a zero for the *y* coordinate (we'll deal with *y* in a bit).

Select *t* from the Variable Tool Panel.

> Set the boundaries for *t* to 0 and 2

Hit play.

What happened to the position of the point?

Does *t* show up on the graph? How?

2. *y* as a function of *t*

The height of the ball also changes with respect to time.
Recall that the function for the height a falling object is

$$y = \tfrac{1}{2}gt^2 + v_0 t + h_0$$

g is the acceleration of gravity, or −32 feet per second squared on earth.

v_0 the initial vertical velocity of the object.

h_0 is the initial height of the object.

Complete the formula for the height of the football:

y = _____

Create point B.

> Constrain its coordinates, but this time type 0 for the *x* coordinate.
> Then type a comma,
> Then type the expression you've written in the blank.

>> Remember to use * for multiply, and use ^2 for an exponent of two.

Hit play.

What happens to point B?

3. (x, y) as a function of t

Of course, the football is only at one point, not at two. To realistically represent the position of the football, create point C.

Constrain the x coordinate as you did for point A.

Constrain the y coordinate as you did for point B.

Hit play.

Describe what you see.

Where is the ball after 1.5 seconds?

How does t appear on the graph?

4. What path is point C traveling?

The function describing the position of a point (x, y) in terms of a third variable, t, is called a **parametric function**:

f(t) = (x, y).

t is the control variable.

the point (x, y) is the output of the function.

In other words, one input variable, t, produces two output variables, x and y, as a coordinate pair. Of course, that means that each output variable needs its own rule.

Click on the **Draw function** icon

Select Parametric for the type.

After X=, type 20*t

After Y=, type −16*t^2+30*t

Start at 0

End at 2

Click OK

Hit Play.

What do you see? Where is the ball (x and y) after 1.5 seconds?

The ball has traveled _____ horizontally from the point where it was kicked and is

_____ feet above the ground.

5. Modify your Geometry Expressions drawing to solve these problems:

   a. A cannonball is fired from a cannon resting on the ground. Its horizontal velocity is 125 feet per second, and its initial vertical velocity is 60 feet per second.

   How long does it take for the cannonball to hit the ground?

   How far is it from the base of the cannon to the crater the cannonball makes in the ground when it lands?

   b. Jackie shoots a basketball from the free-throw line. She releases the ball from a point 6 feet above the ground. Its horizontal velocity is 14 feet per second, and its initial vertical velocity is 20 feet per second.

   How far is the ball from Jackie when it reaches its maximum height?

   What is the maximum height of the ball?

c. In 1990, Randy Byrnes set the world record in the shot put with a throw of 75.85 feet. A competitor releases the shot from a height of 6 feet, with a horizontal velocity of 35 feet per second, and an initial vertical velocity of 32 feet per second.

Will the competitor set a new world record?

6. Summary:

A parametric function describes both _____ and _____ in terms of a third

variable called the _____.

_____ is the control variable for a parametric function.

On an x-y graph, the value of t

_____.

Lesson 3: Go Speed Racer!

Learning Objectives

One of the main ideas of the previous lesson is that the control variable *t* does not appear on the "static" graph of a parametric function. *t* shows its influence through the speed and direction that the point moves along the path.

Math Objectives

- Use Parametric notation.
- Interpret the effect that T has on the graph as motion.

Technology Objectives

- Use Geometry Expressions to demonstrate motion on a parametric graph.

Math Prerequisites

- Function notation.
- Knowledge of parametric functions, as demonstrated in Lesson 2.
- Some knowledge of circular function would be helpful, but not necessary.

Technology Prerequisites

- No special prerequisites beyond what has been learned in this unit so far.

Materials

- Computer with Geometry Expressions.

Overview for the Teacher

1. Though all of the functions for the first part of the lesson involve sin(x), only a cursory understanding of sin(x) is necessary for the lesson – it's just more fun to watch the points move along a sinusoidal curve.

2. Students may run into trouble with constraints if they place the point on the curve, or on the axis. Just have them delete their point and create a new one that is in "white space."

   a) After constraining the point to (T, sin(T)), the point snaps to the sine curve they have drawn. Animating the point will cause it to move along the curve.

   b) Point B will also move along the curve, but at twice the speed.

   c) Changing the parameters of point B to (T, sin(2*T)) moves the point off the curve – its path has half the period as sin(T).

   d) Point B moves along the curve at the same speed as Point A, but from right to left. Both points meet at (0,0).

   e) (T^2, sin(T^2)): x is ≥ 0, and the speed changes.

      (sin(T), sin(sin(T))) causes the point to oscillate.

      There are endless possibilities. The point here is that there is a many-to-one correspondence between parametric functions and their static x-y graphs.

      Parameter T appears on the graph through the movement of the point.

3. Part three of the lesson emphasizes the point that the graph does not tell the whole story. It's a demonstration and requires no written response.

4. The vertical line test does not indicate whether a parametric graph is a function because the parameter T does not appear on the graph.

5. The parametric function for part five describes a circle. If students neglected to change the mode to radians earlier in the lesson, they will need to do so at this point. Also, if they neglect to lock the variable T, its start and end value may change unpredictably.

   If students are still editing the function when they hit play, the old function is used.

   To make the point travel twice as fast, change the function to $\begin{pmatrix} X = \cos(2T) \\ Y = \sin(2T) \end{pmatrix}$

To make the point travel clockwise, students are likely to try $\begin{pmatrix} X = \cos(-T) \\ Y = \sin(-T) \end{pmatrix}$, based on their

results in part 2. If they look closely, they will see that Geometry Expressions has used

trigonometric identities to simplify this to: $\begin{pmatrix} X = \cos(T) \\ Y = -\sin(T) \end{pmatrix}$.

To start at the top and move clockwise: $\begin{pmatrix} X = \sin(T) \\ Y = \cos(T) \end{pmatrix}$ Some students will work with phase

shifts and negatives signs, which can lead into a nice discussion on trig identities.

To double the radius of the circle, change the function to $\begin{pmatrix} X = 2\cos(T) \\ Y = 2\sin(T) \end{pmatrix}$. Some students

will double the circle from the previous qustion, yielding $\begin{pmatrix} X = 2\sin(T) \\ Y = 2\cos(T) \end{pmatrix}$ or some other

variant.

All of the graphs represent functions with respect to T, though their graphs do not pass the vertical line test. The vertical line test is only applicable for functions where x is the control variable.

6. Summary:

A parametric function is a function where the control variable is T and the dependent variables are x and y.
Another name for the control variable is the parameter.
The control variable appears on the graph as motion.
The vertical line test does NOT show whether a parametric curve is a function.

Extension:

One possible solution:

Second hand: $\begin{pmatrix} x = 3\sin(T) \\ y = 3\cos(T) \end{pmatrix}$

Minute hand: $\begin{pmatrix} x = 2\sin\left(\frac{T}{60}\right) \\ y = 2\cos\left(\frac{T}{60}\right) \end{pmatrix}$

Hour hand: $\begin{pmatrix} x = \sin\left(\frac{T}{3600}\right) \\ y = \cos\left(\frac{T}{3600}\right) \end{pmatrix}$

The domain of T will need to be increased to about 377 (120π) to see the entire graph.

The minimum value, maximum value, animation duration in the Variable Tool Panel will need to be adjusted. Note that the maximum for animation duration is 60 seconds.

Student Worksheets

Student worksheets follow.

Name: _____

Date: _____

Go Speed Racer

The location of a point on a plane can be expressed with two separate functions, one for the x coordinate and one for the y coordinate. This combination of functions is called a "parametric function." A third variable, usually T, is the control variable for both functions. T is called "the parameter."

In this lesson, we'll see what happens if you make changes to the parameter, while keeping the functions otherwise the same.

1. Setting up

   Open a new file in Geometry Expressions. Turn on the axis by clicking the axis button.
   Click on Edit on the menu bar, and select Preferences.
   Click Math, and make sure that angle mode is radians.

   Draw function.
   Choose **Cartesian** for **Type.**
   Type in sin(x)

   As you do the lesson, you may wish to zoom in or out to see more of the graph. Use Scale

   Up and Scale Down icons on the top icon bar.

2. What parametric function will travel along this path?

   a. **Draw** Point A, not on the curve.
      Constrain its coordinates to (T, sin(T))
      What happened to the point?

      In the Variables Tool Panel (see diagram 1).
         Select T.
         Set its minimum value to −8 and its
      maximum value to 8.
         Set the animation duration to 10.
         Click on the lock icon.

      Hit Play.

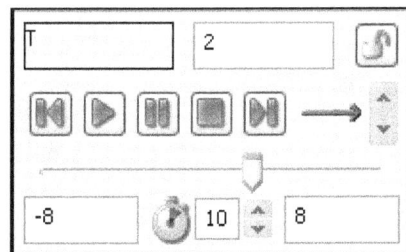

Diagram 1

What do you observe?

b. Draw another point, Point B.
   Constrain its coordinates to (2*T, sin(2*T)).

   Hit Play.
   As you observe the two points, what is the same?

   What is different?

c. Change the coordinates of Point B to (T, sin(2*T)).
   Hit Play.

   What is the same?

   What is different?

d. Change the coordinates of Point B to (-T, sin(-T)).
   What do you think will happen?

   Hit Play.
   What did happen?

e. Try changing the coordinates of Point B to (T^2, sin(T^2)).

   Try changing the coordinates of Point B to (sin(T), sin(sin(T))).

   Try changing the coordinates of Point B to something else that will have the same path as Point A.

   How many different parametric functions share the same set of points as y = sin(x)?

How do they differ?

How does the parameter T appear on the graph?

3. Graphs of parametric functions don't tell the whole story.
    Create a new file in Geometry Expressions.
    Make sure the axis is turned on and that angles are measured in radians.

    Click on **Draw Function.**
        Change the **Type** to **Parametric.**
        Type in X = 2*T
        And Y = sin(2*T)
        Click Enter.

    The graph shows all the points in the graph *but not their speed or direction.*
    To see those features, follow these steps:
        Click on **Draw Point**
        Click on the graph.

        Click on the select icon.
        Press SHIFT and select both the point and the curve.
        Click on **Constrain Point proportional along curve.**
        Type in the Parametric variable T.
        Click on play.

    The graph of a parametric function does not tell the whole story. You also need to describe the motion of the point on the graph.

4. On two of the last examples in part 2, points are repeated more than once.
    Since a parabola is a function, it passes the vertical line test. That's because every value of *x* corresponds with exactly one value of *y*.

    Values of *y* can correspond with more than one value of *x* – that's ok, because *x* doesn't depend on *y*; *y* depends on *x*.

    You can use the vertical line test on a Cartesian (x-y) graph because it *shows* how each *x* value corresponds with just one *y* value.

    For parametric functions, *x* and *y* both depend on *T*. They do not depend on each other.

Will the vertical line test (checking how x corresponds with y) work for parametric functions?

Why or why not?

5. Can you control speed and direction?

    Open a new file in Geometry Expressions.
    Draw a parametric function:
        X = cos (T)
        Y = sin (T)

    Constrain a point proportional to the curve. Name the constraint *t*.
    Select T from the Variable Tool Panel.
        Set its minimum value to 0.
        Set its maximum value to 6.28 (that's about 2π).
        Lock the variable.

    Animate the point.

    Describe the path of the point.

    You can edit the equations for the function by double-clicking on them.
    Try to change the equations so that the point goes around twice as fast.

    Record your parametric function here:

    Change the equations so the point goes clockwise, and record your parametric function here:

    Change the equation so that the point starts at the top of the circle and moves clockwise. Record your function here:

Change the equations so that the circle's radius is doubled. Record your function here:

Do the graphs represent functions with respect to T?

Do the graphs pass the vertical line test?

6. Summary:

A parametric function is a function where the control variable is _____ and the dependent variables are _____ and ____.

Another name for the control variable is the _____.

The control variable appears on the graph as _____

The _____ test does NOT show whether a parametric curve is a function.

Extension: Create the Parametric function for a clock.

Create three parametric functions.

The point on one of the functions will travel like the second hand on a clock.

The point on the second function will travel like the minute hand on a clock.

The point on the third function will travel like the hour hand on a clock.

Try to set up the Variable Tool Panel so that the hands move at the correct speeds.

Lesson 4: Parametric Problems

Learning Objectives

The purpose of this lesson is to apply parametric functions to solving problems in context. Students are expected to solve the problems approximately with technology. Solving the problems algebraically would be a beneficial follow-on activity, but is not included in this lesson.

It is recommended that these problems be done collaboratively, and that brainstorming is encouraged rather than offering a particular method.

Hints are included, but may be withheld at teacher discretion.

Math Objectives

- Model problems in context with parametric functions.

- Problem solve collaboratively in groups.

Technology Objectives

- Use Geometry Expressions to model physical behavior.

Math Prerequisites

- Distance = rate * time and related linear models.

- Formulas for falling/dropped projectile motion.

- Parametric equations for a circle, as learned in this unit.

- Familiarity with parametric equations, as learned in this unit.

Technology Prerequisites

- Geometry Expressions skills, as learned in this unit.

Materials

- Computers with Geometry Expressions.

Overview for the Teacher

Students should use this time to problem solve rather than use a prescribed algorithm. There are hints provided at the end of the worksheet. You may wish to withhold the hints or only use a subset, depending on the level of your students.

Most of the problems include breaking an initial velocity down into horizontal and vertical components. It is not the intent of this lesson to describe this process, as that may not fall nicely into your sequencing. Instead, "fill-in-the-parameter" type functions are included in the hints portion. You may wish to go into this topic in depth before you begin the lesson, or you may wish to leave this topic in a "mysterious state" until a more appropriate time.

Answers provided here are approximate, since it is not the intent of this lesson to use algebraic techniques to obtain exact answers, although that may be a useful follow-on activity.

1. Angles between 42° and 54° will go over the goalposts. Diagrams 1a and 1b show graphs and formulae.

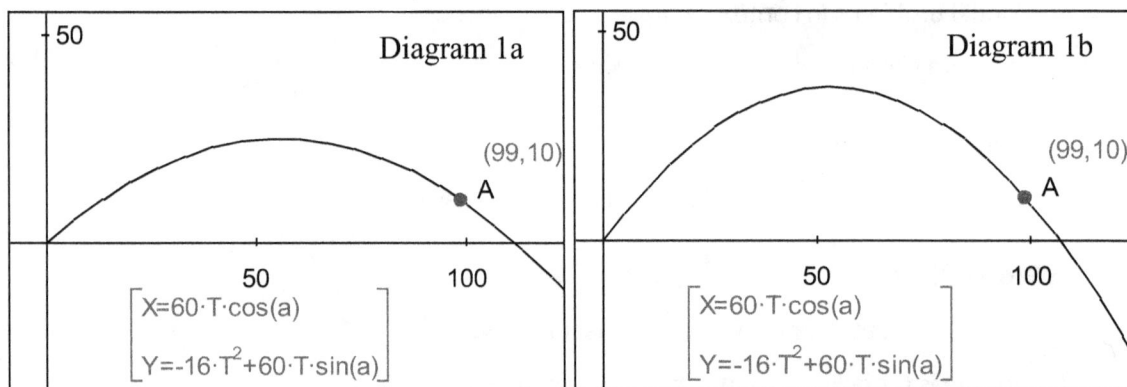

2. The pirate ship should aim at either 15° or 75°. 15° would probably be more effective since it is more of a "direct hit." Diagrams 2a and 2b show the sketch and parametric function.

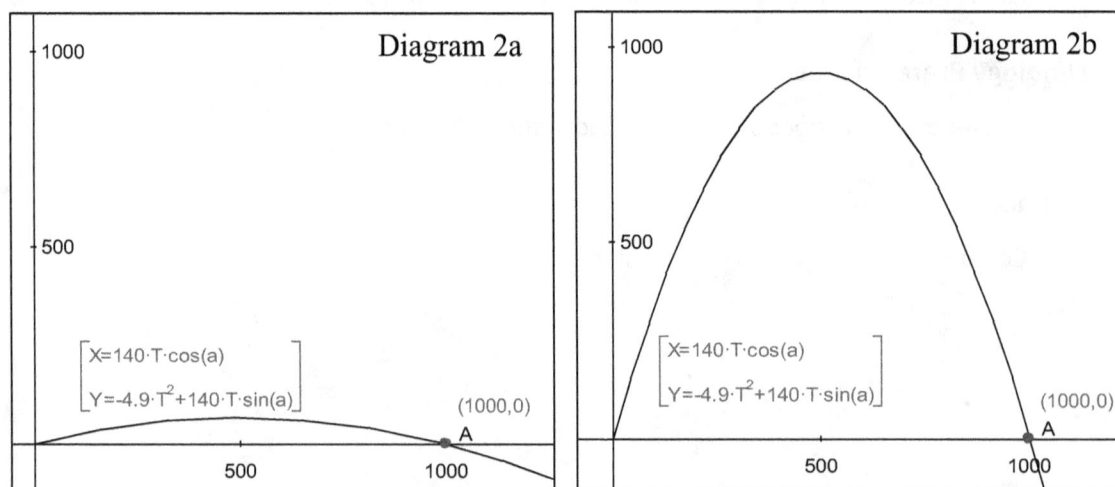

3. The maximum range of the cannon is about 2000 meters (at an angle of 45°). See Diagram 3.

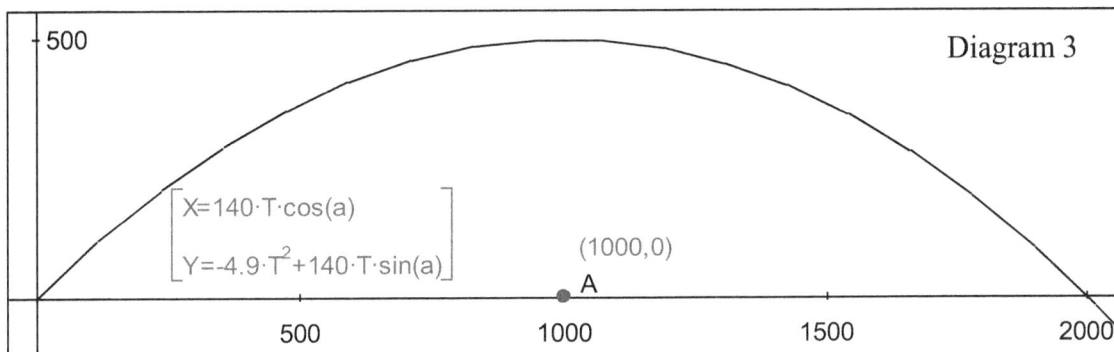

Diagram 3

$$\begin{bmatrix} X=140 \cdot T \cdot \cos(a) \\ Y=-4.9 \cdot T^2 + 140 \cdot T \cdot \sin(a) \end{bmatrix}$$

(1000,0)

A

4. The two cars do not collide. See Diagram 4.

5. The gunner may fire at either 1.43 seconds or at 86 seconds. The second answer would allow too much time for course change or other interference. See Diagrams 5a and 5b.

    Some versions of Geometry Expressions include a bug that will affect this problem. If students are using Draw Function, their angle measures of 38 degrees will be mistranslated. They can work around the bug by clicking on the equation, and re-typing 38.

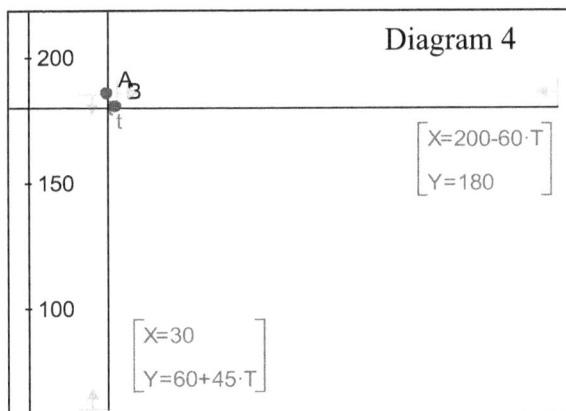

Diagram 4

$$\begin{bmatrix} X=200-60 \cdot T \\ Y=180 \end{bmatrix}$$

$$\begin{bmatrix} X=30 \\ Y=60+45 \cdot T \end{bmatrix}$$

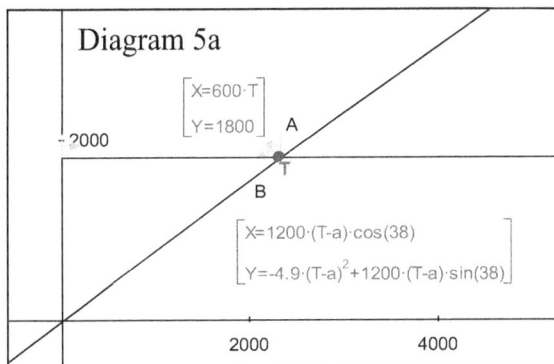

Diagram 5a

$$\begin{bmatrix} X=600 \cdot T \\ Y=1800 \end{bmatrix}$$

A
T
B

$$\begin{bmatrix} X=1200 \cdot (T-a) \cdot \cos(38) \\ Y=-4.9 \cdot (T-a)^2 + 1200 \cdot (T-a) \cdot \sin(38) \end{bmatrix}$$

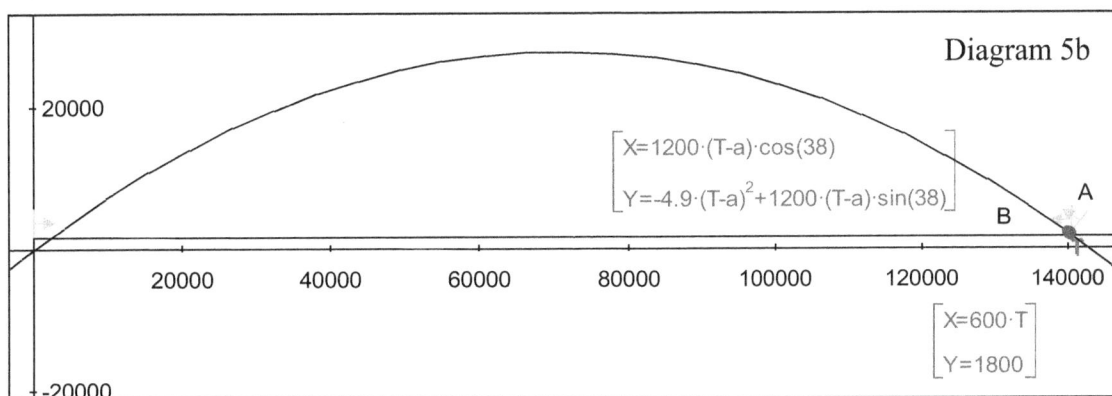

Diagram 5b

$$\begin{bmatrix} X=1200 \cdot (T-a) \cdot \cos(38) \\ Y=-4.9 \cdot (T-a)^2 + 1200 \cdot (T-a) \cdot \sin(38) \end{bmatrix}$$

A
B

$$\begin{bmatrix} X=600 \cdot T \\ Y=1800 \end{bmatrix}$$

6. This problem involves some inductive reasoning. Answers are as follows:

Jogger: 360° or one revolution.
Runner: 180° and 360°.
Biker: 120°, 240°, and 360°.

Mathematician: $\dfrac{360°}{n-1}, \; 2\left(\dfrac{360°}{n-1}\right), \; 3\left(\dfrac{360°}{n-1}\right) \dots (n-1)\left(\dfrac{360°}{n-1}\right)$

See Diagram 6 for a sketch.

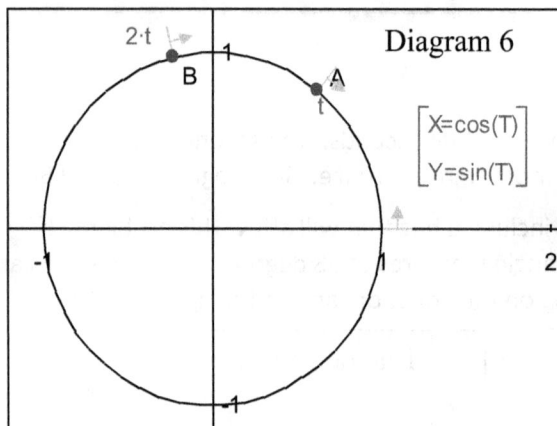

Diagram 6

$$\begin{bmatrix} X=\cos(T) \\ Y=\sin(T) \end{bmatrix}$$

Student Worksheets

Student worksheets follow.

Parametric Problems

Use Geometry Expressions and parametric functions to solve these problems.
Record your parametric function and a sketch along with your solution

Hints for some of the problems are at the end of the handout.

1. A football player is attempting a 33-yard field goal. He can kick the football with a maximum initial velocity of 60 feet per second (See the hint section if you need help breaking this down into horizontal and vertical velocities!). If the goal posts are 10 feet above the ground, at about what range of angles must he kick the ball if he is to clear the goal posts?

2. A pirate ship is firing its cannon at a giant squid. The squid is floating motionless, almost tauntingly, 1000 meters away. Cannonballs fly with a velocity of 140 m/sec. Find two cannon angles that will lead to a direct hit. Which of the two angles do you think would be more effective?

3. What is the maximum range of the cannon in problem 2? Why?

4. A car is heading north from a point with coordinates (30, 60) at a speed of 45 miles per hour. A second car is heading west from a point with coordinates (200, 180) at a speed of 60 miles per hour. Do the cars collide?

5. An enemy spy plane is spotted 1800 meters directly over an anti-aircraft gun. The plane is traveling at 600 meters per second. The gunner has time to rotate the gun turret, but no time to change the angle of the barrel, which is set at 38 degrees. The missile will travel with an initial velocity of 1200 meters per second. How long must the gunner wait before firing the missile, if he is going to shoot down the plane? This problem has two solutions. Why would you disregard the second answer?

6. A person is walking one lap around a circular track. Another person is jogging at twice the speed of the walker. How many degrees around the track does the walker turn before the jogger passes him? A runner is traveling at three times the speed of the walker. At what points does the runner pass the walker? A biker is traveling at four times the speed of the walker – when does the biker pass the walker? A theoretical mathematician is traveling at n times the speed of the walker. How many times, and at which angles does the mathematician pass to the walker in one lap for the walker?

Hints

1. The parametric function for a falling or thrown object subject to gravity is:

$$\begin{pmatrix} x = v_0(\cos\theta)t + x_0 \\ y = \dfrac{g}{2}t^2 + v_0(\sin\theta)t + y_0 \end{pmatrix}$$

   The object is traveling at an initial velocity of v_0 at an angle of θ with the horizontal.

   g is equal to -9.8 meters per second squared or -32 feet per second squared.

   (x_0, y_0) is the initial position of the object.

   Make sure the angle measure is in degrees.

   Select Edit from the menu bar and click on Preferences.

   Click on the Math icon on the left.

   Choose Degrees for angle mode.

   After you type in your equation, Geometry Expressions will sometimes insist that you meant radians! If this is the case, click on the equations, and re-type them.

   Create a point and constrain it to (99, 10). This will mark the low point of the goalposts.

   Change θ so the parabola goes above the point.

2. Read carefully – this problem is in meters. Use g = -9.8 meters per second squared. You can modify your Geometry Expressions drawing from problem 1 to solve this problem.

3. By changing θ, what is the largest x intercept you can get? This is the solution.

4. Your point of view on this problem is that you are looking down at the ground (like a map). If a point is moving due North, South, East, or West, then one of its equations will be a constant. South and West will have negative velocities.

5. Your point of view on this problem is from the ground, a distance away from the anti-aircraft gun. The plane is flying from the left to the right.
   You can delay the missile by using $(t - k)$ instead of t for the parameter. Alternatively, you can try constraining a point proportional to the curve, and then typing $(t - k)$ for the constraint. Modify k to change the time you wait before firing.

6. Remember the parametric equation for a circle from the last lesson. What did you do to get different speeds?

Unit 2: Conics and Loci

Primary Mathematical Goals

- Students will learn the concept of a "locus."

- Students will define circles, ellipses, parabolas, and hyperbolas in geometric and algebraic terms.

- Students will define conics with envelope curves.

- Students will compare parabolas, hyperbolas, and ellipses in terms of focus, directrix, and eccentricity.

Overview

Many traditional studies of conic sections center on the implicit formulas and how they can be used to quickly sketch graphs. These skills are quickly becoming outdated as graphing technologies evolve. Geometric definitions of conic sections are often treated as a separate topic, sometimes even in a separate course.

The intent of this unit is to look at conic sections through three lenses:
> Definitions of conic sections as a locus of points
> Parametric equations of conic sections, in terms of trigonometric functions
> Implicit equations of conic sections

The main goal of this unit is to make connections between these three representations of the conic sections. At the end of the unit, students will have geometric definitions, parametric equations, and implicit equations for circles, ellipses and hyperbolas, as well as a geometric definition and implicit equation for parabolas.

Outline

Lesson 1: Introducing Loci   *Time Required*: 60 – 75 min.

- The concept of a locus of points is introduced. Circles, angle bisectors, and perpendicular bisectors are among the loci that are studied. While interesting in its own right, most of this lesson gets students acquainted with the features of Geometry Expressions that they will need throughout the unit.

Lesson 2: Loci: Circles   *Time Required*: 90 – 120 min.

- The definition of a circle as a locus of points is presented. Connections are made to the unit circle, the Pythagorean Theorem, and to the distance formula.

Lesson 3: The Ellipse   *Time Required*: 90 – 120 min.

- An ellipse is constructed using the geometric definition. Students use Geometry Expressions' symbolic capabilities to generate the parametric and implicit equations.

Lesson 4: Envelope Curves   *Time Required*: 75 – 100 min.

- The concept of a locus of lines and the resulting envelope curve is used to re-define the ellipse. This definition is extended to introduce hyperbolas.

Lesson 5: Inside-Out Ellipses   *Time Required*: 90 – 120 min.

- Lesson 4 is used as a jumping-off point for this lesson on hyperbolas. The geometric definition, parametric equation, and implicit equation are developed through comparison with ellipses.

Lesson 6: Eccentricity and Parabolas   *Time Required*: 60 – 75 min.

- Parabolas are introduced as the locus of points equidistant from a line and a point. The concept of eccentricity is presented to expand this definition to include hyperbolas and ellipses.

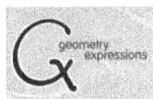

Lesson 1: Introducing Loci

Learning Objectives

This lesson is the introduction to conic sections. Conic sections (parabolas, hyperbolas, ellipses, circles) can all be described as a locus of points. In this lesson, students will become re-acquainted with some familiar loci, and with how Geometry Expressions can be used to explore loci.

Math Objectives

- Learn the concept of a "locus."

Technology Objectives

- Create loci with Geometry Expressions

Math Prerequisites

- Parallel lines, perpendicular bisectors and angle bisectors
- Parametric Functions

Technology Prerequisites

- None

Materials

- Computers with Geometry Expressions

Overview for the Teacher

1. Part one introduces the students to the software, and to the idea of a locus.

   Watch to see if students are changing the value of a – it should remain constant. Otherwise, they might be led to believe that the locus is a family of concentric circles. The Variables Tool Panel can be used to lock variable a. Click on the entry for a, and then click on the lock icon.

   Note that θ measures the rotation of the segment counter-clockwise from the horizontal. However, if the student draws the segment from A to B, the A is placed at the origin. If the segment is drawn from B to A, then B is placed at the origin.

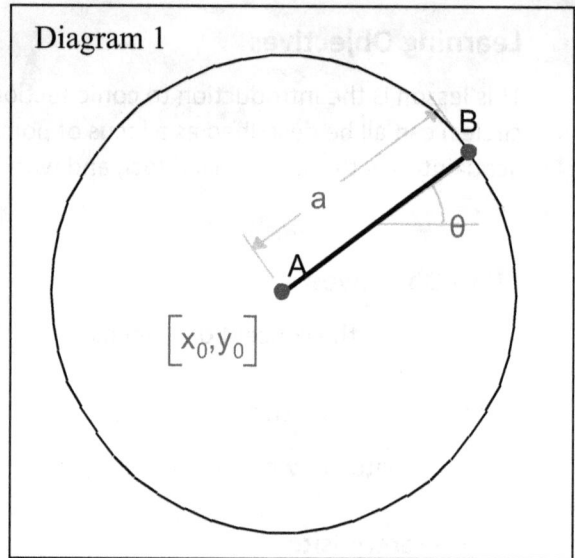

Diagram 1

$[x_0, y_0]$

   The result is a circle with center A and radius a, as seen in Diagram 1.

2. The difficult part of question 2 is making a case for creating point A. To create a locus of points, Geometry Expressions requires a Parametric Variable, which seems artificial in this case.

   Geometry Expressions will not look for the other parallel line. It uses the position of P as a cue to determine which points to draw. The result is shown in Diagram 2.

   Students can get the other half of the locus by duplicating their work on the other side of the line, or by using the Reflection tool in the Construct Tool Panel.

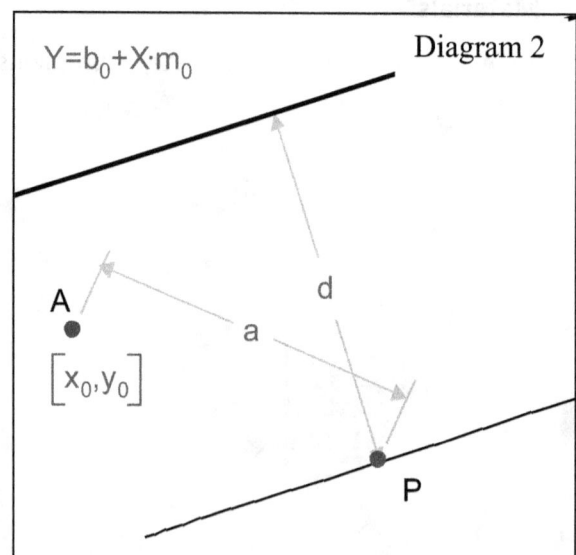

$Y = b_0 + X \cdot m_0$

Diagram 2

$[x_0, y_0]$

The desired result is two lines parallel to the original line, each a distance of *d* from the original line.

3. Students are encouraged to deduce how to use Geometry Expressions more autonomously as the lessons progress.
If students put the point outside of the parallel lines, it will move one of the lines so that they coincide.

It is especially important to lock variable *d* for this part, since that will keep the lines a constant distance apart – they won't wiggle around.

The locus of points equidistant from two parallel lines is a parallel line exactly half-way between the two lines.

For intersecting lines, students may try to create an additional constraint for their parametric variable. Most likely, they will get a message box telling them that they have too many constraints. Remind them that they can use distance *d* as the parametric variable for their locus.

The locus of points equidistant from intersecting lines bisects the angle formed by the lines.

4. Students are expected to create this locus from start to finish. The result is the perpendicular bisector of the line segment.

5. In summary, a locus is a collection of points meeting a given set of criteria. The locus can be described geometrically, for example though distances from fixed points or shapes.

Diagram 3

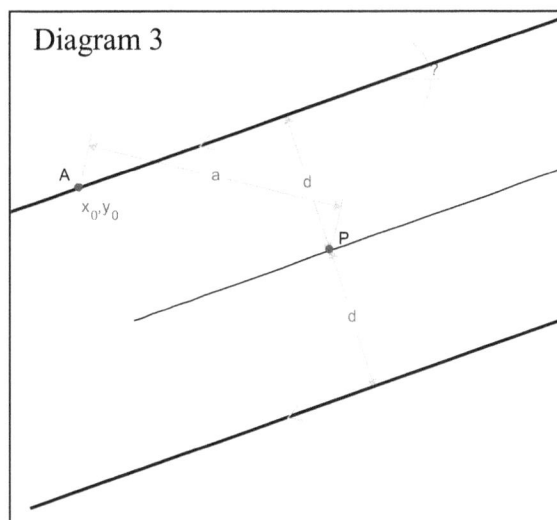

Diagram 4

$Y = b_1 + X \cdot m_1$

$Y = b_0 + X \cdot m_0$

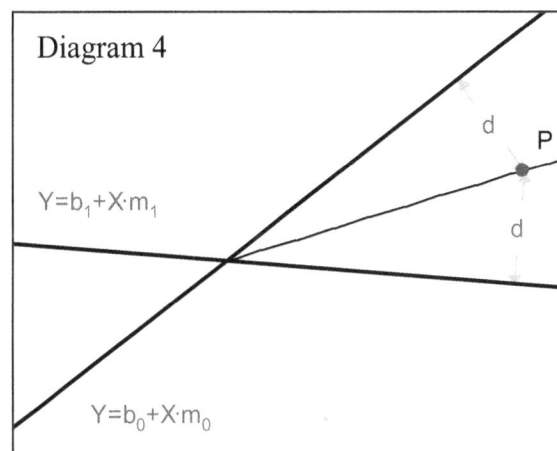

Diagram 5

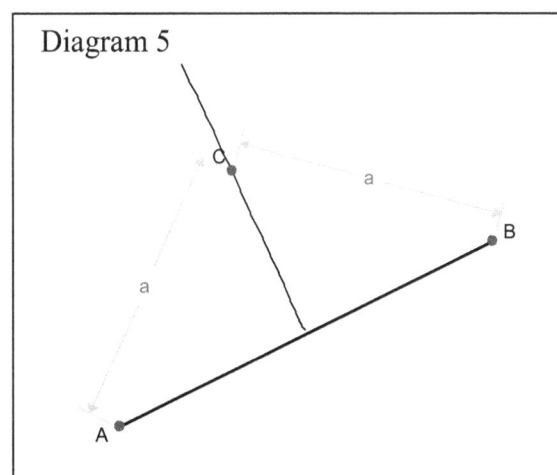

The next lesson in this unit will examine the circle locus more closely. It will create a link between the geometric construction and algebraic equations of conics. Both parametric and implicit forms will be examined.

Student Worksheets

Student worksheets follow.

Name: _____

Date: _____

Loci

A Locus is a set of all the points that meet a particular description.
"What are all the points a certain distance from a fixed point?"
"If I tie my dog to a stake, what are the boundaries that he can't cross?"

1. Find the locus of points equidistant from a fixed point (like all the places a dog can go if it is tied to a stake).

    Open a new drawing with Geometry Expressions

    Draw point A.

    Constrain its location to (x_0, y_0).

    Click on the point, and then click on

    Draw point B.

    Constrain the distance between the points to *a*.
    Hold the shift key and select each point. Then

    click on

Sketch the locus of points equidistant from a fixed point.

    Draw a line segment connecting the two points.

    Constrain the direction of the line segment to θ.

    Choose θ in the Variables Tool Panel. Use the slider bar to see the locations that point B can occupy.

    Construct the locus of the points, using θ for the parametric variable.
    Select point B

    Click on **Construct locus**
    Choose θ for the Parametric Variable.

    How would you describe this locus?

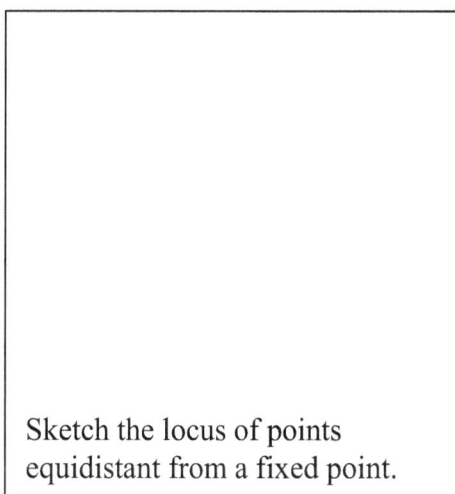

2. What is the locus of points equidistant from a line (perhaps the path a dog creates along the edge of a fence)?

Open a new drawing with Geometry Expressions.

Draw a line.

Constrain the line's implicit equation. You can leave the equation at its default setting.

Draw point P so that it is not on the line. **Constrain** its distance from the line to be *d*.

You can lock variables by selecting them in the Variables Tool Panel, and then clicking the lock icon. Lock variable d.

Drag point P to see its locus – all of the points that are *d* units from the line.

Geometry Expressions will draw the locus, but it wants a point of reference that *changes*.

We'll ask it to draw each point that is *d* units from the line, but at all the *different* distances from some fixed point. So, we need to draw a fixed point.

Draw point A anywhere.

Constrain its coordinates.

Constrain the distance from A to P to *t*. This constraint is called **a parametric variable.**

To see the graph of the locus, Select point P.

Construct its **locus**.
Use *t* for its Parametric Variable.

Are there any points that are distance *d* away from the line, but that weren't drawn? Computers aren't always able to do a complete job, based on your description. Unlock variable *d,* and try dragging P again and see what happens. Are there any other locations possible for P? Sketch them in your diagram.

Sketch the locus of points equidistant from a line.

Sketch the locus of points equidistant from two parallel lines.

3. What is the locus of points equidistant from two lines?

There are two cases.

Case number 1: Two parallel lines:

Draw a line. **Constrain** one of the line's equations. **Draw** a second line, and **constrain** the two lines to be parallel

Draw point P between the two lines. **Constrain** the distance from P to each line to be *d*.

Locus of points equidistant from the endpoints of a line segment.

Think about how you created a parametric variable for your locus in part 2. Create a parametric variable, and create the locus. What is the locus of points equidistant from two lines, if the lines are parallel?

Case number 2: Two intersecting lines:

Draw intersecting lines, and **constrain** both of their equations.

Draw point P, constrained to be distance *d* from each line.

For parallel lines, the distance from the point to the lines was always the same. For intersecting lines, the distance changes. That means you can use *d* as your parametric variable when you create the locus.

Describe the locus in your words.

Locus of points equidistant from two intersecting lines.

What is the special name for this locus?

4. What is the locus of points equidistant from the endpoints of a line segment?

    Use the line segment tool to **draw** a line segment.
    Explore the locus of points equidistant from the endpoints of the line segment.

    Describe the locus in your words.

    What is the special name for this locus?

5. Summarize what you have learned in this lesson.

    What is the definition of a locus?

    What are some different ways that a locus can be described?

Lesson 2: The Circle

Learning Objectives

Students are now acquainted with the idea of "locus," and how Geometry Expressions can be used to explore loci. Now, they will look more closely at the circle, defined as the locus of points on a plane equidistant from a fixed point. In particular, they will find the parametric and implicit equations of circles.

The approach is to make sense of the parametric equation in terms of the unit circle, and the implicit equation in terms of the distance formula. Review of these two concepts may be beneficial before the start of the lesson.

Math Objectives

- Learn or re-enforce the general parametric and implicit equations of a circle.
- Connect translations and dilations to the general equations.
- Connect center and radius to the general equations.

Technology Objectives

- Use Geometry Expressions to find equations of curves.
- Use Geometry Expressions to translate and dilate figures.

Math Prerequisites

- Distance formula
- Algebra, including factoring parts of an expression and completing the square.
- Unit Circle
- Pythagorean identity
- Translations and dilations
- Parametric functions vs. implicit equations

Technology Prerequisites

- Knowledge of Geometry Expressions from previous lessons.

Materials

- Computers with Geometry Expressions.

Overview for the Teacher

1. Question one gets the students back into using Geometry Expressions to find the locus of a point. Diagram 1 exhibits typical results.

    A common error is to use *r* (the default) as the parametric variable instead of changing it to θ.

    Students are asked why they didn't just use the circle tool. Expected results are that the definition of a circle as a locus of points was reinforced.

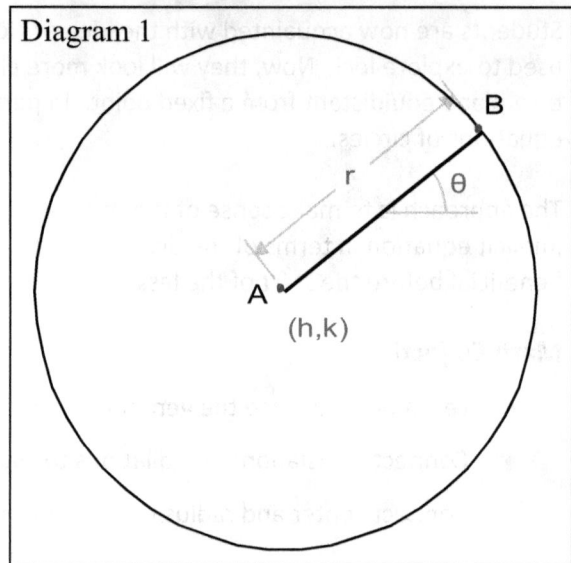

Diagram 1

2. In Diagram 2 you can see the general parametric equation of a circle, where *d* is the radius and (h, k) is the center. If students are getting numerical constants, they are calculating the *real* equation instead of the *symbolic* equation. Help them to select the *symbolic* tab and try again.

    After changing the constraints so that the center is (0,0) and the radius is 1, the parametric equation will be

$$\begin{pmatrix} X = \cos(\theta) \\ Y = \sin(\theta) \end{pmatrix}$$

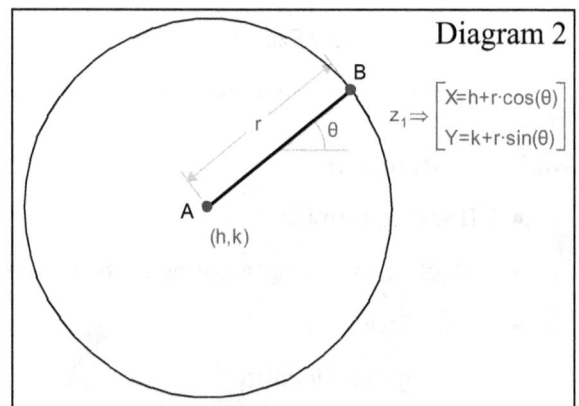

Diagram 2

$$z_1 \Rightarrow \begin{bmatrix} X=h+r\cdot\cos(\theta) \\ Y=k+r\cdot\sin(\theta) \end{bmatrix}$$

If students are not getting this, they may be changing the value of the *d* in the Variable Tool Panel, rather than changing the constraint itself to 1.

3. Diagram 3 shows the results of finding the implicit equation for the circle. The simplification process involves some grouping and simple factoring. The end result is the distance formula:

$r^2 = (X-h)^2 + (Y-k)^2$, which is

frequently written $r^2 = (x-h)^2 + (y-k)^2$

Changing the center to (0,0) and the radius to 1 results in $-1 + X^2 + Y^2 = 0$ or

$X^2 + Y^2 = 1$

Diagram 3

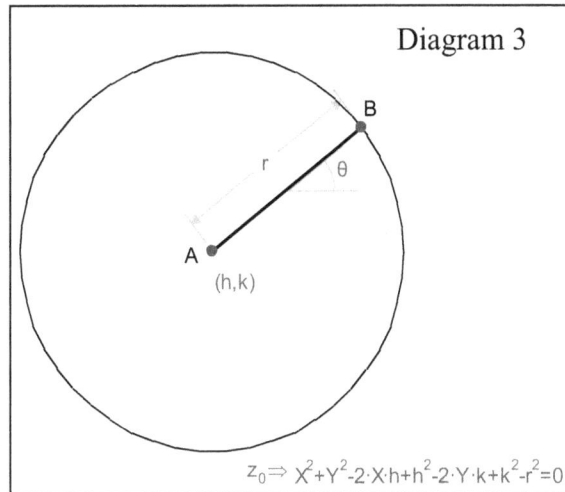

$z_0 \Rightarrow X^2 + Y^2 - 2 \cdot X \cdot h + h^2 - 2 \cdot Y \cdot k + k^2 - r^2 = 0$

Exceptional students will begin making connections between the equation of a circle, the distance formula, and the Pythagorean Theorem at this point. Most will likely have been exposed to these connections in a previous course. You may wish to highlight these relationships at this time.

4. The result is the Pythagorean Identity: $\cos^2 \theta + \sin^2 \theta = 1$

5. Translating and dilating the unit circle will restore us to the general form. Note that the center will be (u_0, v_0) instead of (h, k). If students find this confusing, they can change the variables used in the vector to (h, k).

Encourage students to change the vector by dragging D, especially if they cannot see the whole picture on the screen.

If your students are not comfortable with completing the square, then the generalization of the implicit equation may be a demonstration.

Diagram 5

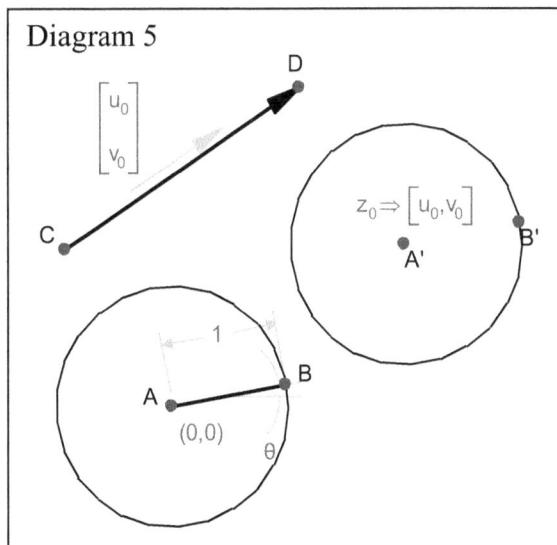

6. If students get incorrect results for their dilations, check to see that they have chosen the center of the translated circle as their center of dilation. If they choose some other point, their dilated circle will not be concentric with the translated circle, and their equations will be wrong. See diagram 6 for typical correct results.

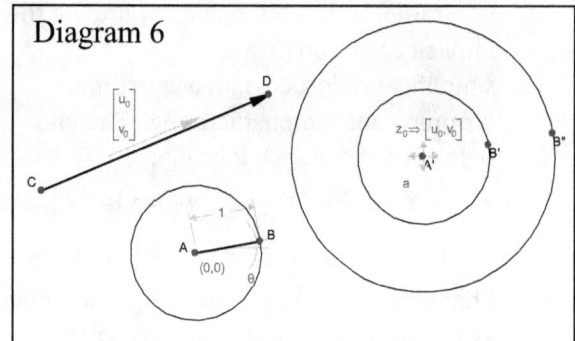

Diagram 6

7. In summary:
    The general parametric form of the equation of a circle is

$$\begin{pmatrix} X = h + r\cos\theta \\ Y = h + r\sin\theta \end{pmatrix}$$

(x_0, y_0) is the center of the circle and d is the radius.

The general implicit formula of a circle is
$$r^2 = (X - h)^2 + (Y - k)^2$$
Again, (h, k) is the center of the circle and d is the radius.

It is important that students have these three concepts, as they will be repeated for the rest of the conic sections:

1.   Description of the curve as a locus of points.

2.   Parametric equation of the curve.

3.   Implicit equation of the curve.

Subsequent lessons will search for further attributes of the curve being studied. For example, the study of ellipses will include the focus, major axis and minor axis.

Student Worksheets

Student worksheets follow.

The Circle

In the last lesson, you were introduced to the idea of a "locus." A locus is a collection of points that meet a particular description. For example, the locus of points equidistant from a fixed point is a circle.

1. Reconstruct the locus of points equidistant from a fixed point.

    Open Geometry Expressions.
    Create point A, and constrain its location to (h, k).
    Create point B, and constrain its distance from point A to be r.

    Create line segment $\overline{AB}$ – be sure to draw starting at point A and ending at point B – and constrain its direction to be θ.

    Create the locus of points d units from point A. Use θ as your parameter.

    Sketch the locus of points equidistant from a fixed point.

    You could have just used the circle tool if all you wanted to do was to draw a circle. What did you learn about circles by doing it this way?

2. Often, a locus can be described with equations describing its x and y coordinates. This type of algebraic description is called a Parametric Equation. X and Y are described as functions of a third variable, called a parameter.

    Click on the circle you created in step one.

    Click on Calculate Symbolic Parametric equation.

    Geometry Expressions has just given you the General Parametric Equation of a circle. Write it in the box to the right.

    Write the general parametric equation of a circle

Recall the Unit Circle. How are the coordinates of points on the unit circle defined in terms of cosine and sine?

Change the constraint on the center to (0,0) and change the radius to 1. What does this do to the Parametric equation?

3. You can use CTRL-Z to undo your changes to the constraints on your drawing. Do so repeatedly, until the radius is r and the center is (h, k). Then use Geometry Expressions to create the implicit equation.
Select the circle.

Click on Calculate Implicit equation [x/] .

The result looks confusing at first, but you can clean it up:

Add d^2 to both sides.
Group the X and x_0 terms.
Group the Y and y_0 terms.
What formula is beginning to emerge?

Factor the X and x_0 terms.
Factor the Y and y_0 terms.
The result is the General equation of a circle.

> Simplify the Implicit Equation here, to get the General Equation of a circle.

Change the constraints as you did in part 2. The result is the implicit equation for the unit circle. Write it here:

4. Using the equations for the unit circle, substitute the two parametric equations into the implicit equation. What relationship do you get?

5. We are going to subject our unit circle to some transformations. First, translate the unit circle.

Create a vector.

Constrain the components of the vector to $\begin{pmatrix} u_0 \\ v_0 \end{pmatrix}$.

Select the circle and its center (use shift and click on each).

Click Construct Translation

Click the vector.

What are the coordinates of the center of the translated circle?

Click on the center.

Click on Calculate Symbolic Coordinates.

Find the parametric equation of the translated circle.

Where do the coordinates of the center appear in the equations?

Find the implicit equation of the translated circle.

Use the method of *completing the square* to simplify the implicit equation.

6. Now, dilate the circle that you translated.

Click on the circle.

Click on Construct Dilation.
Click on the center of the translated circle.
Type *r* for your scale factor.

What is the radius of the dilated circle?

Find the parametric equation of the translated/dilated circle.

Translation of a circle

Sketch the translation and dilation of the unit circle.

Where does a, the length of the radius appear in the equations?

Find the implicit equation of the translated/dilated circle.

Simplify with completing the square, as you did in part 5.

7. Summary

The general parametric form of the equation of a circle is:

_____

Where _____ is the center of the circle and _____ is the radius of the circle.

The general implicit form of the equation of a circle is:

_____

Where _____ is the center of the circle and _____ is the radius of the circle.

Lesson 3: The Ellipse

Learning Objectives

In this lesson, students will generalize their knowledge of the circle to the ellipse. The parametric and implicit equations of an ellipse will be generated, as will two important properties of the ellipse: that the sum of the distances from any point to the foci is equal to the major diameter, and the Pythagorean Property of Ellipses.

Math Objectives

- Understand the geometric definition of an ellipse.

- Generate the parametric and implicit equations for an ellipse.

- Locate the foci, given the equation of an ellipse.

- Discover relationships between the parameters of an ellipse.

Technology Objectives

- Use Geometry Expressions to create a more complex locus of points.

- Find evidence for equivalence using Geometry Expressions.

Math Prerequisites

- Pythagorean Theorem

- Translations

- Parametric functions and implicit equations

- Sine and Cosine

Technology Prerequisites

- Knowledge of Geometry Expressions from previous lessons.

Materials

- Computers with Geometry Expressions.

Overview for the Teacher

1. Diagram 1 represents a typical result for question 1. If students are getting semicircles with center at point F1, they are creating a locus in terms of t instead of d. Only half an ellipse is shown because the entire ellipse cannot be generated as a function of d. Points below the foci will have the same distances as points above.

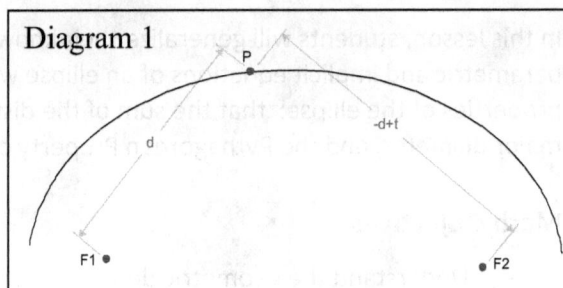

Diagram 1

An ellipse is like a circle in that it is a curve based on distances from fixed points. It is different in that it "has different radii."

2. An appropriate result would be $\begin{pmatrix} X = a\cos\theta \\ Y = b\sin\theta \end{pmatrix}$. Transposing the *sine* and *cosine* functions will have no effect on the final curve. Note that the transposed version is also the "sample" parametric function given by the software. Using the transposed version yields the same results, graphically. It just changes the starting place and direction that the ellipse is graphed.

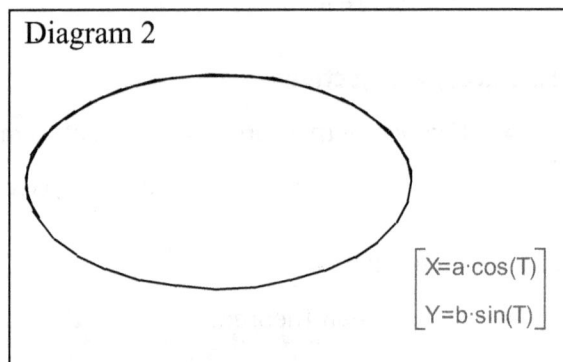

Diagram 2

$$\begin{bmatrix} X = a \cdot \cos(T) \\ Y = b \cdot \sin(T) \end{bmatrix}$$

Encourage students to change values for a and b to get an ellipse that is not just a circle. A sample is shown in Diagram 2

3. Some assumptions are made here about the symmetrical nature of an ellipse. You may wish to explore these assumptions with the class at this time.

Desired solutions:

a. The distance from F1 to P is $2a - m$ or $2c + m$

b. The distance from F2 to P is m

c. The sum is therefore $2a$ or $2c + 2m$.

d. The width of the ellipse is $2a$

e. $t = 2a$. Remind students that they need to type 2*a .

©2009 Saltire Software Incorporated

4. The constraint from F2 to P is $2a - d$. Diagram 3 shows expected results.

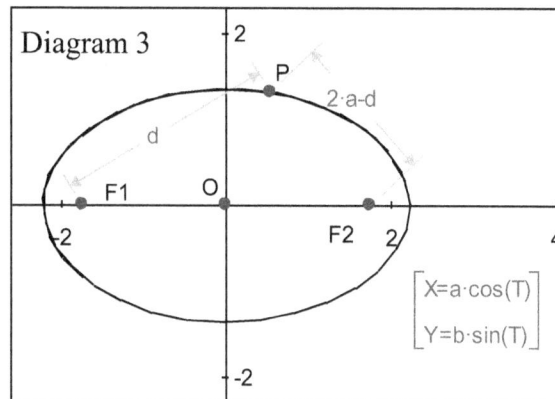

5. Desired solutions:

    a. The sum of the distances is 2a.

    b. The distance from F2 to P is equal to a, since the three points form an isosceles triangle.

    c. Using the Pythagorean theorem,

    $$a^2 = b^2 + c^2, \text{ so } c = \sqrt{a^2 - b^2}$$

    d. Point P will now appear to be on the ellipse, as shown in Diagram 4.

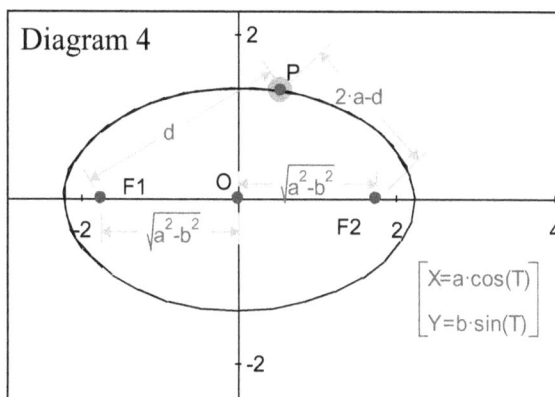

6. In both instances, the implicit equation will be

    $$Y^2 \cdot a^2 + X^2 \cdot b^2 - a^2 \cdot b^2 = 0$$

7. The general parametric function that is generated is $\begin{pmatrix} X = u_0 + a\cos(T) \\ Y = v_0 + b\sin(T) \end{pmatrix}$

The implicit formula is $Y^2 a^2 + X^2 b^2 - a^2 b^2 - 2Xb^2 u_0 + b^2 u_0^2 - 2Ya^2 v_0 + a^2 v_0^2 = 0$

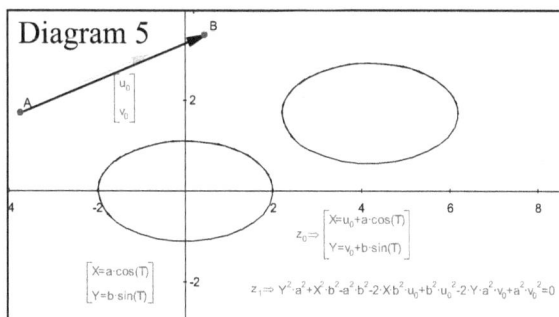

8. Steps are as follows:

$$Y^2a^2 + X^2b^2 - a^2b^2 - 2Xb^2u_0 + b^2u_0{}^2 - 2Ya^2v_0 + a^2v_0{}^2 = 0$$

$$\left(X^2b^2 - 2Xb^2u_0 + b^2u_0{}^2\right) + \left(Y^2a^2 - 2Ya^2v_0 + a^2v_0{}^2\right) = a^2b^2$$

$$b^2\left(X^2 - 2Xu_0 + u_0{}^2\right) + a^2\left(Y^2 - 2Yv_0 + v_0{}^2\right) = a^2b^2$$

$$b^2\left(X - u_0\right)^2 + a^2\left(Y - v_0\right)^2 = a^2b^2$$

$$\frac{b^2\left(X - u_0\right)^2}{a^2b^2} + \frac{a^2\left(Y - v_0\right)^2}{a^2b^2} = \frac{a^2b^2}{a^2b^2}$$

$$\frac{\left(X - u_0\right)^2}{a^2} + \frac{\left(Y - v_0\right)^2}{b^2} = 1$$

9. Results as follows:

   a. If $a = b$, then the result is a circle.

   b. If $a < b$, then the ellipse is taller than it is wide. The foci lie on a vertical line rather than
       a horizontal line. The Pythagorean Property would then be $b^2 = a^2 + c^2$

10. Summary:

The general parametric form of the equation of an ellipse is $\begin{pmatrix} X = u_0 + a\cos(T) \\ Y = v_0 + b\sin(T) \end{pmatrix}$

where (u0, v0) is the center of the ellipse and a and b are the radii of the ellipse.

The general implicit form of the equation of an ellipse is $\dfrac{\left(X - u_0\right)^2}{a^2} + \dfrac{\left(Y - v_0\right)^2}{b^2} = 1$

where (u0, v0) is the center of the ellipse.

If $a > b$, then 2a is the major diameter and 2b is the minor diameter.

If $b < a$, then 2b is the major diameter and 2a is the minor diameter.

The sum of the distances from any point on the ellipse to the two foci is 2a

The distance from the center of the ellipse to either focus follows the equation $c^2 = a^2 + b^2$

Student Worksheets

Student worksheets follow.

Name: _____

Date: _____

The Ellipse

In the last lesson, you found equations for a circle: the locus of points equidistant from a fixed point.

An ellipse is defined as all the points such that the **sum** of the distance from **two** fixed points is a constant. . Each of the two points is called a **focus** (the plural of "focus" is "**foci**").

Diagram 1   P

d   t – d

F1

F2

$D + (t - d) = $ a constant

1. Create an ellipse using the definition above.
    Open a new Geometry Expressions drawing.
    Create two points, and name them F1 and F2.
    Constrain the coordinates of these points.

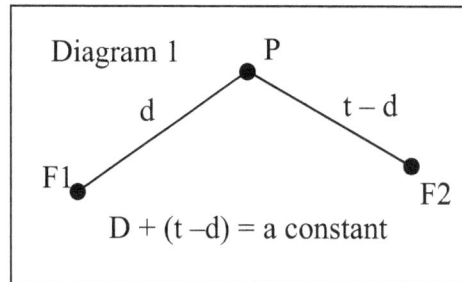

    Create a third point, and name it P.
    Constrain the distance from P to F1 to be *d.*
    Constrain the distance from P to F2 to be *t – d* (see Diagram 1 to understand why!)
    Lock the value of *t* in the Variable Panel, but keep *d* unlocked.

    Find the locus of P with parameter *d.*

    If you drag P around, you can see that the locus forms part of an ellipse.
    To get more of the ellipse:
        Double click on the curve.
        Change the Start Value and End Value for *d*

    The most you can get is half of the ellipse, because the software assumes you know which side P is on – you drew it there! To get the rest of the ellipse:
    Draw a line segment from F1 to F2.
    Select the locus curve.

    Click on **construct reflection** and then click on the segment.
    Some of the ellipse may still be missing.

    How is an ellipse like a circle? How is it different?

Sketch of the ellipse

Before continuing, make sure Geometry Express is set to Radians.
In the Edit Menu
> Select preferences.
> Click on the Math icon at the left.
> Under Math, change Angle Mode to radians.

2. Recall that the general parametric equation for a circle with the center at the origin is $\begin{pmatrix} X = r\cos(T) \\ Y = r\sin(T) \end{pmatrix}$. Predict the general parametric equation for an ellipse.

> Test your prediction with Geometry Expressions
> > Open a new Geometry Expressions drawing.
> > Click on the Function tool in the Draw tool panel.
> > Type in your prediction to see if you are right. Make additional guesses if you need to.

3. The curve you created in part 2 looks like an ellipse, but is it really an ellipse? If it is, we will be able to find its foci, and the constant sum of distances from the foci.

    a. In Diagram 2, how far is it from focus F1 to point P?

    b. How far is it from focus F2 to point P?

    c. What is the sum of distances from P to the two foci?

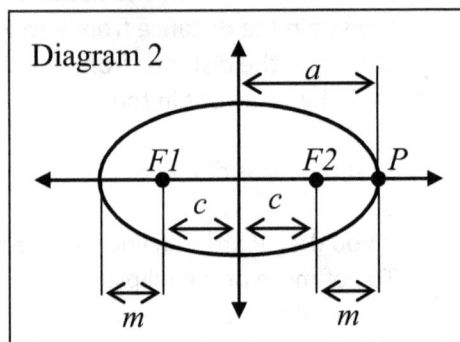

Diagram 2

    d. What is the horizontal width of the ellipse?

    e. Write an expression for your answer to part c, in terms of distance *a*.

4. Open a new Geometry Expressions drawing, and create this parametric function:

$$\begin{pmatrix} X = a * \cos(T) \\ Y = b * \sin(T) \end{pmatrix}$$

Use the Variable Tool Panel to change the values of *a* and *b* so that *a* is greater than *b*. Lock variable *a* and *b*.

Add three points to your Geometry Expressions drawing.
    First, turn on the axes.
    Draw F1 and F2 on the x-axis.
    Draw P so that it is not on either axis, nor is it on the curve.
    Constrain the distance from F1 to P to be *d*.

What should the constraint from F2 to P be? Review the results from part 3 to help you decide.

5. Refer to Diagram 3 to find the positions of F1 and F2. The triangle shown is an isosceles triangle, with P at the vertex.

    a. If P is on the ellipse, what is the sum of the distances from F1 to P and from F2 to P (your solution to 3c)?

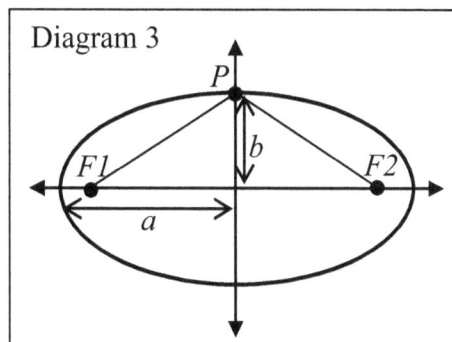

Diagram 3

    b. Given that the triangle is isosceles, what is the distance from F1 to point P?

    c. Use the Pythagorean Theorem to write an expression for the distance from the origin to point F2.

    d. In your Geometry Expressions drawing, constrain the distance from F1 to the origin to your answer to 5c. Do the same for the distance from F2 to the origin. (You will need to draw a point at the origin first). NOTE: If you want to type a square root in to Geometry Expressions, use sqrt, and if you want to type in an exponent, use ^. For example,
    $\sqrt{a^2 + b^2}$ can be typed: sqrt(a^2 + b^2)

    e. Does P appear to fall on the ellipse? Drag it around. Does it stay on the ellipse?

6. If the curve in your Geometry Expressions drawing is truly an ellipse, then its implicit equation will match the locus of point P.
Select the curve, and click on the Calculate Implicit Equation icon.
Now, hide the curve.
   Select the curve.
   Right click on the curve.
   Choose hide.

Create the locus of point P with respect to d, and calculate its implicit equation.

Restore the original curve by clicking on Show All in the View menu.

Are the two implicit equations the same? How do they differ, if at all?

7. How does a translation affect the equation of an ellipse?
Open a new drawing and use Draw Function to create a new ellipse.

Choose Parametric for the type and enter $\begin{pmatrix} X = a\cos(T) \\ Y = b\sin(T) \end{pmatrix}$.

Create a vector, and constrain it to its default values, $\begin{pmatrix} u_0 \\ v_0 \end{pmatrix}$.

Translate the ellipse.
   Select the ellipse
   Click on Construct Translation
   Click on the vector.

Calculate the parametric equation of the new ellipse, and record it in the box.

Calculate the implicit equation of the new ellipse.

General Parametric
Equation of an Ellipse

8. The General form for the implicit equation of an ellipse is $\dfrac{(x - u_0)^2}{a^2} + \dfrac{(y - v_0)^2}{b^2} = 1$. Verify that your implicit equation is equivalent to the general form.

9. In part 6, you found a relationship known as "The Pythagorean Property for Ellipses"
$a^2 = b^2 + c^2$
a is half the horizontal axis of the ellipse
b is half the vertical axis of the ellipse
c is the distance from the center of the ellipse to each focus

a.   What happens if *a* = b?

Diagram 4

b.   Is it possible for *a* < *b*? How would you need to modify the Pythagorean Property for the ellipse in Diagram 4?

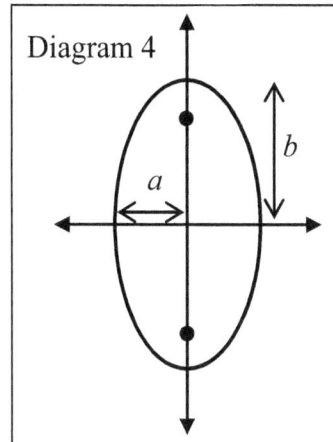

In any ellipse, the larger of 2*a* and 2*b* is called **the major axis.** The smaller of 2*a* and 2*b* is called **the minor axis.** If *a* > *b*, then the foci lie on a horizontal line. If *a* < *b*, then the foci lie on a vertical line. If *a* = *b*, then the ellipse is actually a circle.

10. Summary:

The general parametric form of the equation of an ellipse is:

_____

where _____ is the center of the ellipse, is the horizontal radius of the ellipse, and ____and

_____ are the radii of the ellipse.

The general implicit form of the equation of an ellipse is:

_____

where _____ is the center of the ellipse.

If _____, then ____ is the major diameter and ____ is the minor diameter.

If _____, then ____ is the major diameter and ____ is the minor diameter.

The sum of the distances from any point on the ellipse to the two foci is _____.

The distance from the center of the ellipse to either focus follows the equation _____.

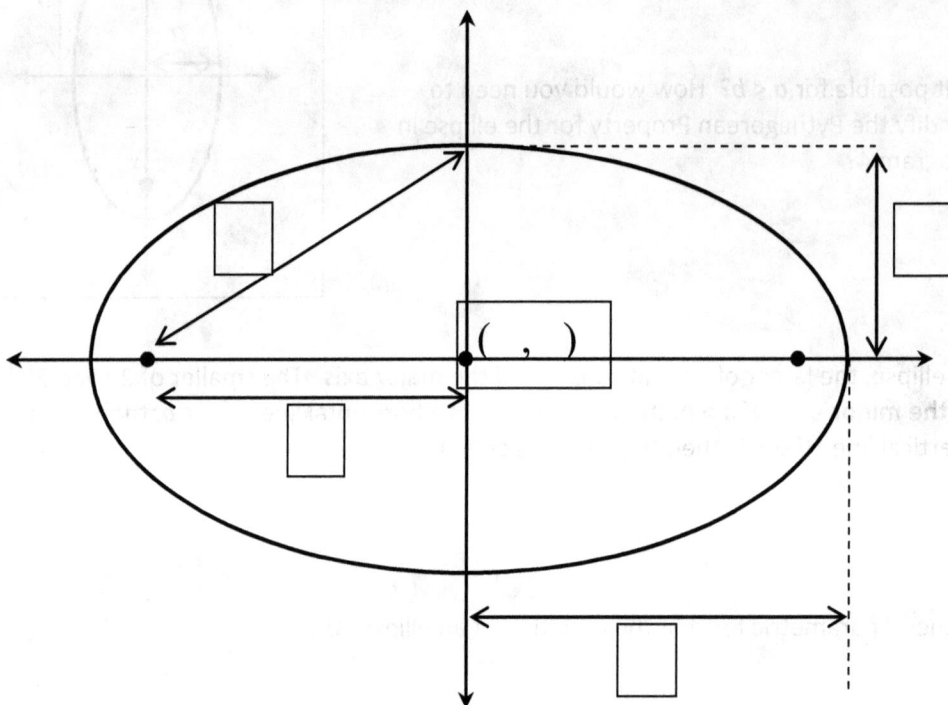

(,)

Conics and Loci Lesson 4: Conics and Envelope Curves
Level: Pre-Calculus
Time required: 90 minutes

Lesson 4: Conics and Envelope Curves

Learning Objectives

Students begin by looking at an envelope curve that generates an ellipse. The curve is modified to become a hyperbola, thereby introducing that concept.

Math Objectives

- Extend the idea of a locus of points to a locus of lines.

- Find a definition of the ellipse as an envelope curve.

- Find definitions for hyperbola, first as a locus of lines, and then as an envelope curve.

Technology Objectives

- Use Geometry Expressions to create an envelope curve

- Use Geometry Expressions as an aide to creating geometric proof.

Math Prerequisites

- Two-column triangle congruence proofs

Technology Prerequisites

- Skills with Geometry Expressions, as developed in previous lessons.

Materials

- Computer with Geometry Expressions

Conics and Loci Lesson 4: Conics and Envelope Curves
Level: Pre-Calculus
Time required: 90 minutes

Overview for the Teacher

1. The purpose of the first part of the lesson is to become familiar with the envelope curve created by a locus of lines, and the related Gemetry Expressions capabilities. It is important that students follow directions closely for part 1. Some will attempt to draw in all of the lines, copying the diagram in the student master. This is not really helpful for the rest of the lesson.

   An error that is more difficult to diagonose may occur if students have their axis turned on, use the origin for point A, and place points B and C on the axes. When they try to set points M and N proportional along the curve, they will run into an error. That is because the software knows M and N are on the axis, but does not know they are bounded to the segments. The easy fix is probably to turn off the axis and start over.

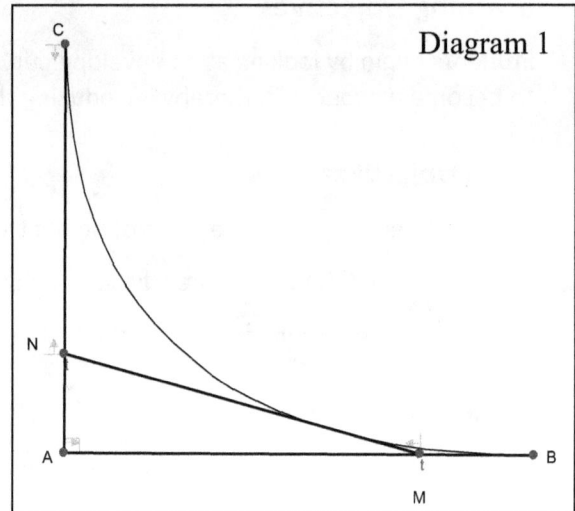

Diagram 1

2. Diagram 2 shows the expected result. If the envelope curve does not show up, it is most likely because the student has chosen the wrong parametric variable for the locus. The parameter should be θ. Another possibility is that the domain for θ is incorrect. By default, it is assigned values between 0 and 6.28 (2π).

   Generating the locus of point E with parameter θ will result in the same curve.

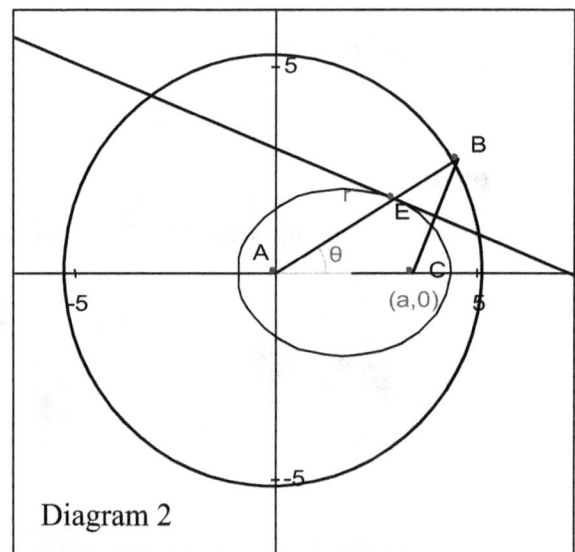

Diagram 2

Conics and Loci Lesson 4: Conics and Envelope Curves
Level: Pre-Calculus
Time required: 90 minutes

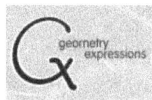

3. You may choose whether to do this proof as a class, guide them through it, or let individuals or groups complete the proof, depending on what is right for your class.

Statement	Reason
$\overline{ED} \perp \overline{BC}$	Given
$\angle EDC \cong \angle EDB$	Right angles are congruent
$\overline{BD} \cong \overline{CD}$	Definition of bisector
$\overline{ED} \cong \overline{ED}$	Reflexive property of congruence
$\triangle EBD \cong \triangle ECD$	Side-Angle-Side
$\overline{EB} \cong \overline{EC}$	CPCTC
AE + EB = AE + EC	Additive property of equality
AE + EB = r, the radius of the circle	Given
AE + EC = r, the radius of the circle	Substitution
AE + AC is a constant	Definition of a radius
The locus is an ellipse	Definition of an ellipse

4. Most of the assumptions that Geometry Expression makes when it is asked to "use assumptions" have to do with whether one value is greater than another. In this case, the drawing indicates $a < r$, and that is the basic assumption that is made. If this is not assumed, Geometry Expressions uses absolute values in place of the assumption, and the symbolics are not as fully simplified.

Watch students to make sure they type the expression in correctly, and with the right subscripts. When testing the lesson with current software, the expression resolved to r. It is possible that, depending on how the drawing is constructed, the expression will not be completely simplified. In that case, help students to finish the simplification process themselves.

Conics and Loci Lesson 4: Conics and Envelope Curves
Level: Pre-Calculus
Time required: 90 minutes

Diagram 3 shows typical student results.

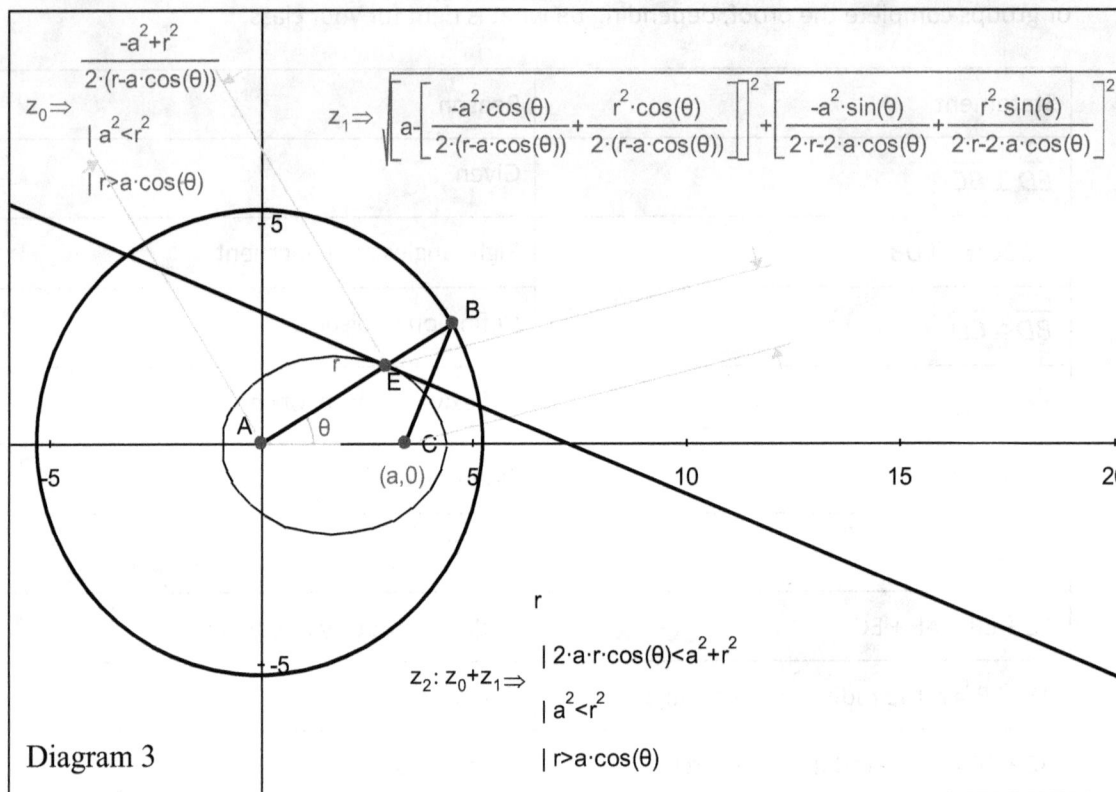

$z_0 \Rightarrow$ $\dfrac{-a^2+r^2}{2\cdot(r-a\cdot\cos(\theta))}$

$| \ a^2 < r^2$

$| \ r > a\cdot\cos(\theta)$

$z_1 \Rightarrow \sqrt{\left[a-\left[\dfrac{-a^2\cdot\cos(\theta)}{2\cdot(r-a\cdot\cos(\theta))}+\dfrac{r^2\cdot\cos(\theta)}{2\cdot(r-a\cdot\cos(\theta))}\right]\right]^2+\left[\dfrac{-a^2\cdot\sin(\theta)}{2\cdot r-2\cdot a\cdot\cos(\theta)}+\dfrac{r^2\cdot\sin(\theta)}{2\cdot r-2\cdot a\cdot\cos(\theta)}\right]^2}$

$z_2 : z_0 + z_1 \Rightarrow$
$| \ 2\cdot a\cdot r\cdot\cos(\theta) < a^2+r^2$
$| \ a^2 < r^2$
$| \ r > a\cdot\cos(\theta)$

Diagram 3

5. The reason that students need to delete the calculated distances first is that Geometry Expressions is assuming that $r > a$, and we are about to change that.

   It is best if point E appears on the branch of the hyperbola that is near point C. Students can toggle the value of θ to make this happen. Diagram 4 includes the expression for the sum of the two lengths.

   The successful student will try calculating the difference of the two lengths. Some will reverse the order, resulting in $-r$ instead of r. This corresponds to E being on the other branch of the hyperbola.

Conics and Loci Lesson 4: Conics and Envelope Curves
Level: Pre-Calculus
Time required: 90 minutes

Diagram 4

$$z_2 \Rightarrow \dfrac{a^2-r^2}{2\cdot(-r+a\cdot\cos(\theta))}\ \big|\ a^2>r^2\ \big|\ r<a\cdot\cos(\theta)$$

$$z_0 \Rightarrow \sqrt{\left[a-\left[\dfrac{-a^2\cdot\cos(\theta)}{2\cdot(r-a\cdot\cos(\theta))}+\dfrac{r^2\cdot\cos(\theta)}{2\cdot(r-a\cdot\cos(\theta))}\right]\right]^2+\left[\dfrac{-a^2\cdot\sin(\theta)}{2\cdot r-2\cdot a\cdot\cos(\theta)}+\dfrac{r^2\cdot\sin(\theta)}{2\cdot r-2\cdot a\cdot\cos(\theta)}\right]^2}$$

$$z_1 : z_2+z_0 \Rightarrow \dfrac{2\cdot a^2-2\cdot a\cdot r\cdot\cos(\theta)}{2\cdot(-r+a\cdot\cos(\theta))}\ \big|\ 2\cdot a\cdot r\cdot\cos(\theta)<a^2+r^2\ \big|\ a^2>r^2$$

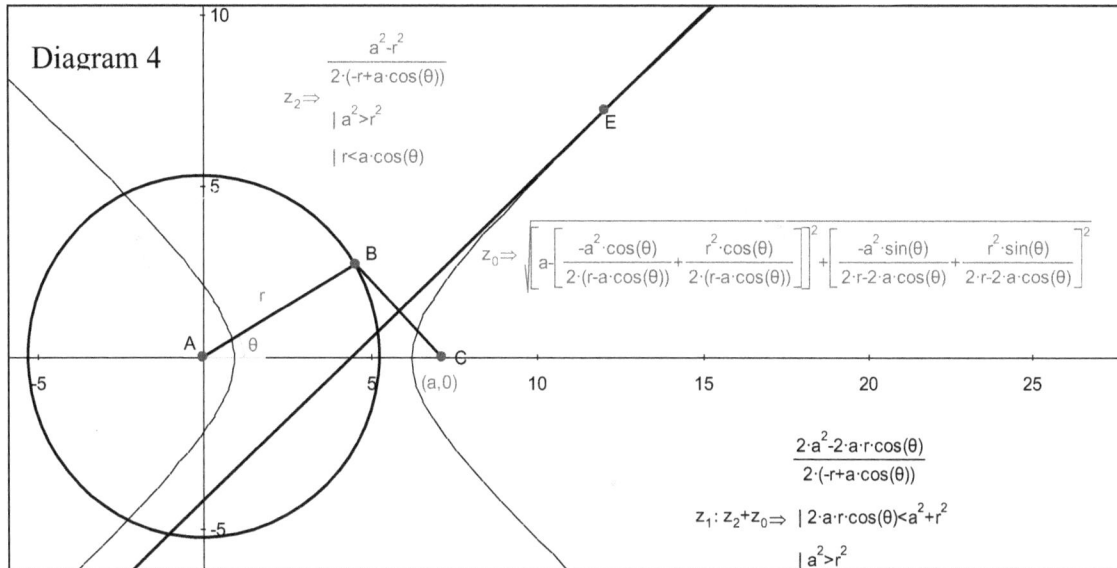

6. The result for part 6 is a circle with center (0,0) and radius $r/2$. While this is not a dramatic or important result, it does satisfy the need mathematicians have for "completeness."

    The next lesson in this series will develop parametric and implicit functions for hyperbolas. The implicit function is quite similar to the ellipse, but the parametric formulas are a little more surprising.

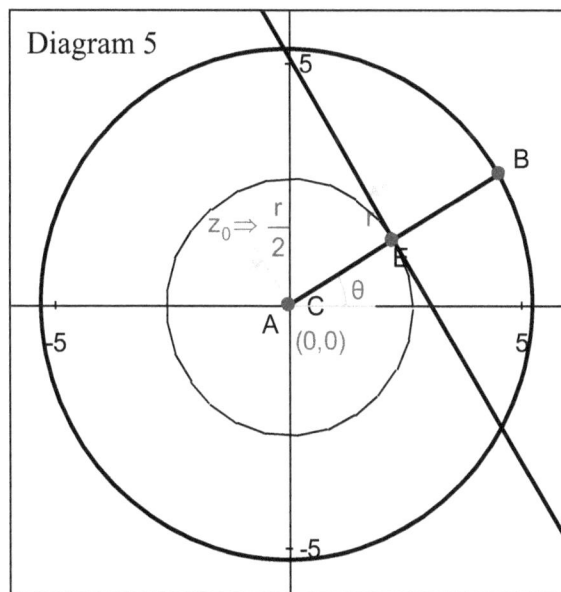

Diagram 5

$$z_0 \Rightarrow \dfrac{r}{2}$$

Student Worksheets

Student worksheets follow.

Name: _____

Date: _____

Conics and Envelope Curves

If you've ever done a "string art" project, then you are familiar with the concept of an envelope curve. You can see a curve begin to form where two adjacent lines intersect. The envelope curve is the locus of these intersection points.

Diagram 1

1. Create the locus of points that are intersections of the lines shown in Diagram 1. Note that Geometry Expressions will not draw all of the lines. It will only draw the curve formed by all the lines.

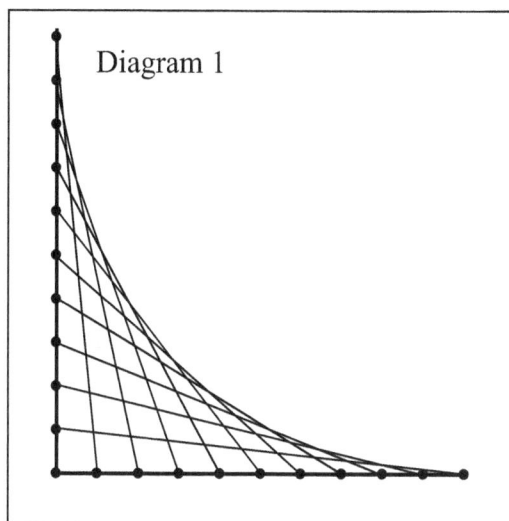

   Open Geometry Expressions.

   Create line segment $\overline{AB}$, and place point M on the line segment.

   Select the point and the segment, and click on the Constrain Point proportional along curve tool. Leave the constraint to be *t*.

   Choose *t* from Variable tool panel, and animate it. Watch the motion of the point along the segment.

   Create a second line segment, $\overline{CA}$. Make sure to start the line segment with point C and to end it with point A, or your point will move in the wrong direction.

   Place point N on segment $\overline{CA}$. Select the point and the segment and click on Constrain point proportional along curve. Change the constraint to be *t*.

   Press play again and watch the points. Make sure that one point is moving towards A while the other is moving away from A.

   Create segment $\overline{MN}$. Play the animation again.

   Now, select $\overline{MN}$ and click on the Construct Locus tool.

   NOTE: the resulting curve is the "envelope" of the lines. The actual locus of lines is the "string" in the string art.

Animate your locus once more. Notice how the line just slides along the curve.

2. Follow these steps to create a second envelope curve.

   Open a new drawing.
   Create a circle with center A at the origin, and point B NOT on the x or y axis. Constrain the radius to be *r*.
   Create radius $\overline{AB}$. Constrain the angle between the radius and the x-axis to be θ.

   Create point C not on the x-axis, in the interior of the circle. Constrain the coordinates of point C to be (a, 0).
   Create segment $\overline{BC}$.

   Construct the perpendicular bisector of $\overline{BC}$ by selecting the segment and then clicking on the Construct Perpendicular bisector tool ◢. You may wish to change the color of this line to help you see it better. Select the line, right click, and choose properties from the menu.

   Animate variable θ. Try to visualize the envelope curve created by the perpendicular bisector. It might be helpful to construct point E at the intersection of the radius and the perpendicular bisector.

   Now, select the perpendicular bisector and click the locus tool. Use θ as the parametric variable. What shape is created? Are you sure?

3. How can we prove that the shape is an ellipse? Refer to diagram 2.

   To show that the shape is an ellipse, prove that AE + CE is equal to a constant.

   Given: $\overrightarrow{ED}$ is the perpendicular bisector of $\overline{BC}$

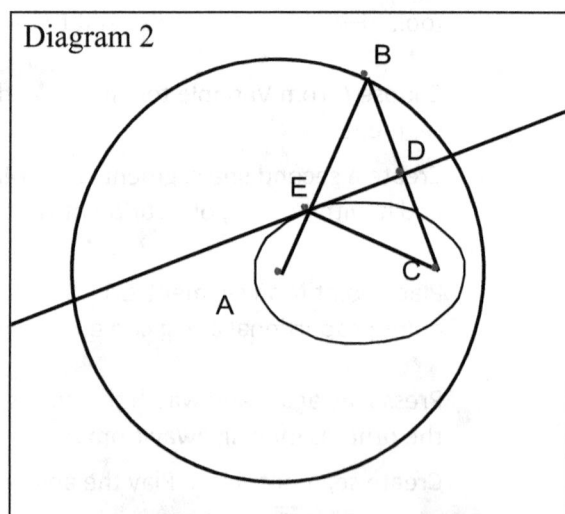

Diagram 2

4. Go back to the Geometry Expressions, and we'll try to confirm our proof.
   If you have not already done so, construct the intersection of the radius and the perpendicular bisector. Select $\overline{AB}$ and the perpendicular bisector, and click on the Construct Intersection tool ◢. Label the point E.

Before continuing, turn on "use assumptions." We are going to allow Geometry Expressions to assume that length r is greater than length a. It will make some calculations simpler, but we'll have to be careful later.

    Click the Edit menu and select Preferences.
    Click on Math
    Under the heading Output, set Use Assumptions to True.

Select A and E and **calculate symbolic** the distance between the points.
Select E and C and **calculate symbolic** the distance between those points.

Both symbolic representations are quite complex. But what, if the shape is an ellipse, what will be true if we add them together?

Click on the Draw Expressions tool and then click on the drawing.
Type z[0]+z[1] and hit Enter.

The result is the first line shown. All of the lines after that are the assumptions being used. What is the result? Does it help prove that the shape is an ellipse? Why?

5.   What happens if point C is outside the circle?

Slowly drag point C until it is outside the circle. If you lose track of point E, change the value of θ with the slider bar in the Variables panel until it appears.

The locus curve that you see now is called a **hyperbola.**

Notice that the calculations for AE, CE, and the sum are still using the same assumptions.
Select each calculation and:
    Right click, and select Properties
    Change Use Assumptions to False.

Then
Select each calculation and:
    Right click, and select Properties
    Change Use Assumptions to True.

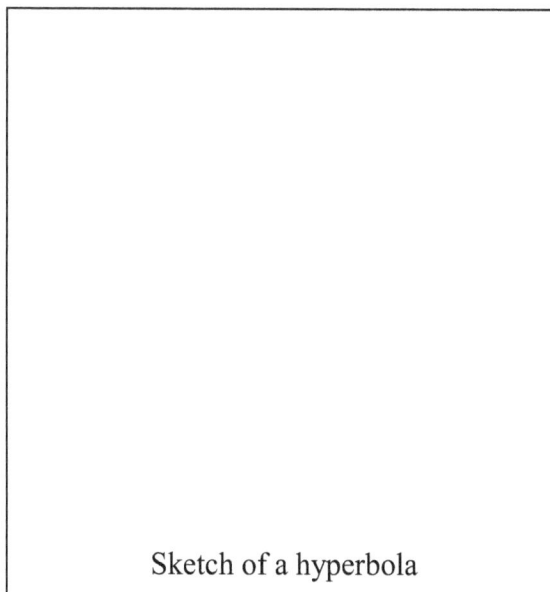

Sketch of a hyperbola

Now Geometry Expressions will assume that $a > r$ which agrees with point C being outside the circle.
Is the sum of the distances still a constant, or is the expression more complex?

Try changing the plus sign to something else. Double click on the expression z_2 to edit it. When the result is a constant, you can complete the definition of a hyperbola:

A hyperbola is the locus of points such that the _____ of the distances from the foci is a constant.

6. What happens if point C is the center of the circle?

   Constrain the coordinates of point C to be (0,0).

   Calculate AE as you did in parts 4 and 5 (CE and AE are the same segment now).

   How would you describe this locus of points?

Conics and Loci Lesson 5: Inside-out Ellipses
Level: Pre-Calculus
Time required: 120 minutes

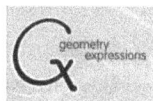

Lesson 5: Inside Out Ellipses

Learning Objectives

What, exactly, is a hyperbola? What are its characteristics and defining equations? How does it relate to the unit circle, and to ellipses? These are the primary objectives of this lesson.

Math Objectives

- Find out what a hyperbola is, and how it is related to an ellipses.

- Learn about characteristics of a hyperbola, particularly its foci and its asymptotes

- Discover General and parametric equations for hyperbolas.

Technology Objectives

- Use Geometry Expressions to draw hyperbolas, using the geometric definition, the parametric equation, and the implicit equation.

- Use Geometry Expressions to connect these three disparate representations of a hyperbola.

Math Prerequisites

- Definitions of secant and tangent, both in terms of right triangles and in terms of sine and cosine.

- Pythagorean identity.

- Solving simple algebra equations.

- Knowledge of circles and ellipses, as provided in earlier lessons of this unit.

Materials

- Computer with Geometry Expressions.

Conics and Loci Lesson 5: Inside-out Ellipses
Level: Pre-Calculus
Time required: 120 minutes

Overview for the Teacher

The student title for this lesson is "Inside-out Ellipses." This presages the next lesson, which is on eccentricity.

The main goal for this lesson is to get a feel for hyperbolas and how they relate to ellipses and circles. Ultimately, the students will have a geometric definition of the hyperbola as a locus of points, and algebraic definitions in parametric form and in implicit form.

1. In part 1, students use the parametric and implicit equations of the unit circle to reproduce the Pythagorean Identity, $\sin^2(\theta) + \cos^2(\theta) = 1$. Diagram 1 shows what students should be seeing.

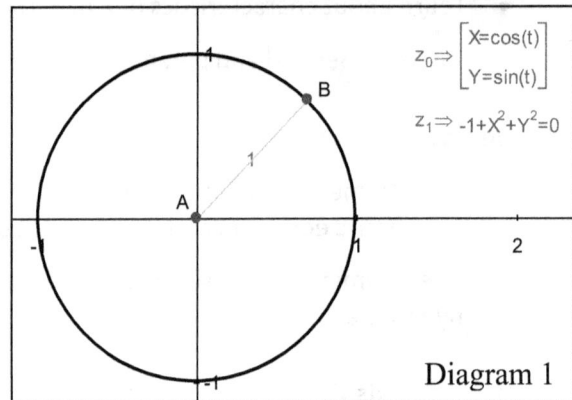

$$z_0 \Rightarrow \begin{bmatrix} X = \cos(t) \\ Y = \sin(t) \end{bmatrix}$$

$$z_1 \Rightarrow -1 + X^2 + Y^2 = 0$$

Diagram 1

2. Here is the sequence for question 2.

$$\cos^2(T) + \sin^2(T) = 1$$
$$\cos^2(T) = 1 - \sin^2(T)$$
$$\frac{\cos^2(T)}{\cos^2(T)} = \frac{1}{\cos^2(T)} - \frac{\sin^2(T)}{\cos^2(T)}$$
$$1 = \sec^2(T) - \tan^2(T)$$

Thus, x will be $\sec(T)$ and y will be $\tan(T)$

Diagram 2 shows the hyperbola and its parametric function.

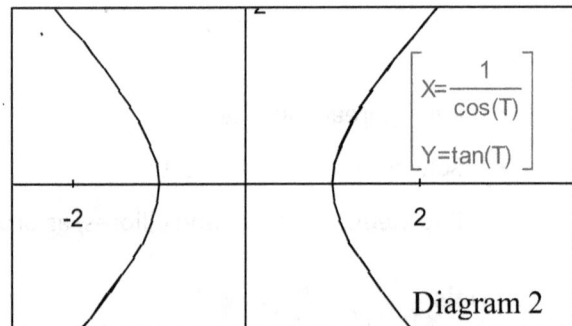

$$\begin{bmatrix} X = \dfrac{1}{\cos(T)} \\ Y = \tan(T) \end{bmatrix}$$

Diagram 2

3. The asymptotes are going to be used to find the foci of this hyperbola. They are also interesting in their own right.

Some students will still have the hyperbola on the same drawing as the unit circle, and others will not. Makes sure they have a both the hyperbola and the unit circle when they proceed to part 4.

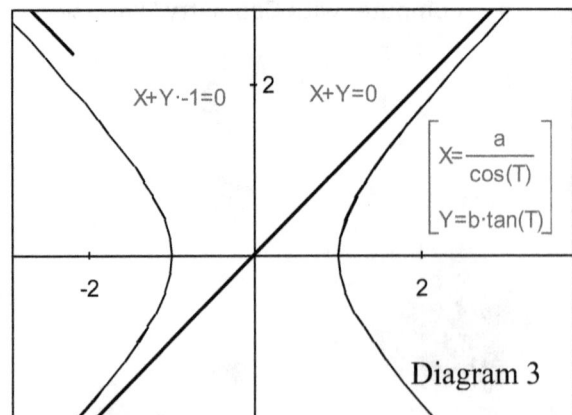

$$X + Y - 1 = 0 \qquad X + Y = 0$$

$$\begin{bmatrix} X = \dfrac{a}{\cos(T)} \\ Y = b \cdot \tan(T) \end{bmatrix}$$

Diagram 3

Conics and Loci Lesson 5: Inside-out Ellipses
Level: Pre-Calculus
Time required: 120 minutes

Diagram 4

4. A small leap of faith is needed in this step, since no proof is offered that the foci can be constructed in this manner. Rather, the foci are used to create a hyperbola that results in a match with this one. Diagram 4 shows student results before all the lines are hidden.

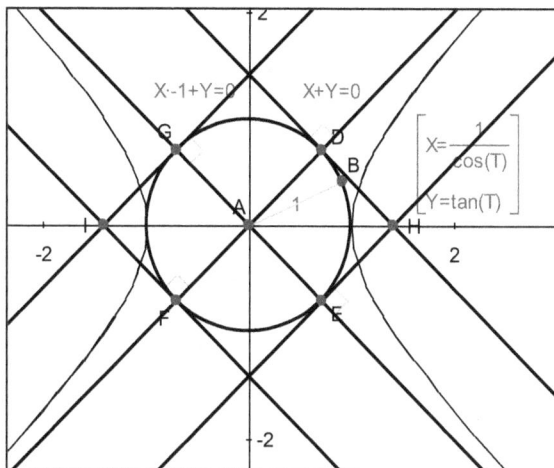

5. The answer to "What is AC − CD" is 2, so the distances from the foci to the point are d and $d + 2$.
   The point falls on the existing hyperbola. The equation for the locus is
   $1 - X^2 + Y^2 = 0$. Some students may need help recognizing this as $X^2 - Y^2 = 1$, our starting point.

6. Changing a stretches the hyperbola horizontally, but since this also moves the foci towards or away from the origin, the hyperbola changes vertically too. Changing b only stretches the hyperbola vertically since the foci are not affected.

The ellipse is tangent to the hyperbola at its vertices.

The line $y = \dfrac{b}{a}x$ is one of the asymptotes for the hyperbola. The other asymptote is $y = -\dfrac{b}{a}x$ You may wish to point out that $\dfrac{b}{a}$ and $-\dfrac{b}{a}$ have opposite signs, but are not reciprocals. Therefore, the asymptotes are not perpendicular

See Diagram 5 for results.

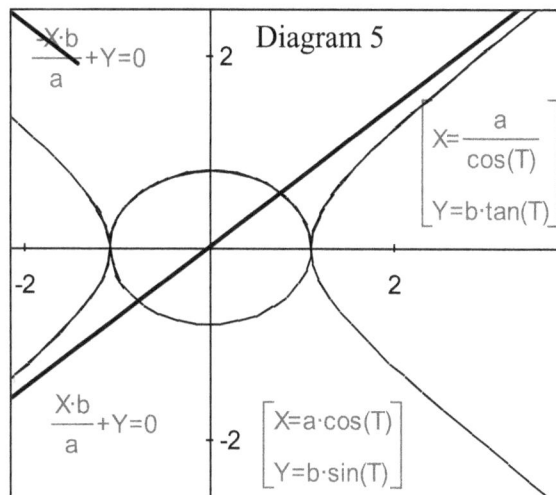

Diagram 5

Conics and Loci Lesson 5: Inside-out Ellipses
Level: Pre-Calculus
Time required: 120 minutes

7. The parametric function for a hyperbola translated by $\begin{pmatrix} u_0 \\ v_0 \end{pmatrix}$ is shown in Diagram 6.

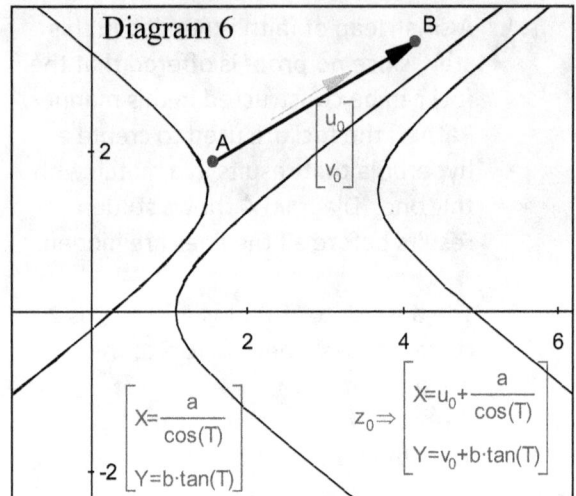

Diagram 6

$$X = \frac{a}{\cos(T)}$$
$$Y = b \cdot \tan(T)$$

$$z_0 \Rightarrow \begin{aligned} X = u_0 + \frac{a}{\cos(T)} \\ Y = v_0 + b \cdot \tan(T) \end{aligned}$$

8. Inductive reasoning is used at this point because it reinforces the roles of the different parameters. You may wish to derive the general implicit equation in class as well.

Start with the trig identity derived earlier: $\sec^2 T - \tan^2 T = 1$.

Solve each of the parametric equations for sec T and tan T, respectively.

Substitute the results into the identity.

Make a few algebraic adjustments, and you are finished.

The general equation for a hyperbola is $\dfrac{(X - u_0)^2}{a^2} - \dfrac{(Y - v_0)^2}{b^2} = 1$

a stretches the hyperbola horizontally
b stretches the hyperbola vertically
u_0 is the horizontal translation
v_0 is the vertical translation

9. Students should graph the implicit function $y = \sqrt{1 + x^2}$. They will see the top half of a hyperbola that opens up, as seen in Diagram 7. The other half of the hyperbola is produces by $y = -\sqrt{1 + x^2}$. A drawing of the parametric form is shown in Diagram 8. Both parts of the hyperbola are drawn because no plus-or-minus is required in the parametric form.

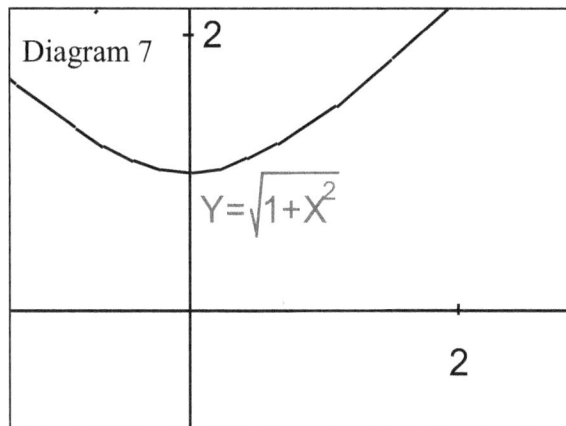

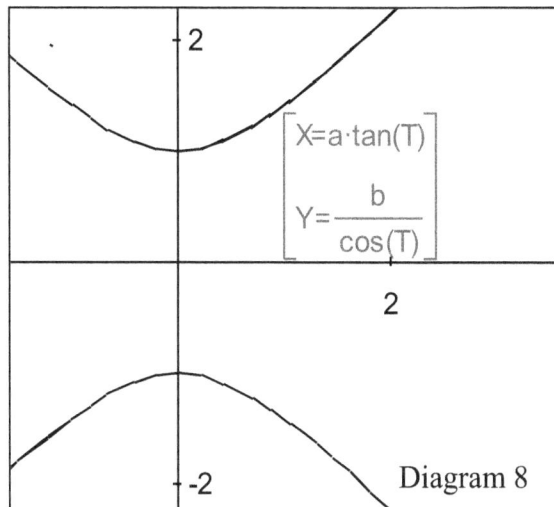

Conics and Loci Lesson 5: Inside-out Ellipses
Level: Pre-Calculus
Time required: 120 minutes

Diagram 7

$$Y=\sqrt{1+X^2}$$

$$\left[X=a\cdot\tan(T) \quad Y=\dfrac{b}{\cos(T)} \right]$$

Diagram 8

10. Summary:

A hyperbola is the locus of points for which the difference of the distances from the foci is a constant.

If a hyperbola opens to the left and to the right:

Its parametric equation is:
$$\begin{pmatrix} X = a\sec(T)+u_0 \\ Y = b\tan(T)+v_0 \end{pmatrix}$$

Its implicit equation is:
$$\frac{(X-u_0)^2}{a^2} - \frac{(Y-v_0)^2}{b^2} = 1$$

If a hyperbola opens up and down:

Its parametric equation is:
$$\begin{pmatrix} X = a\tan(T)+u_0 \\ Y = b\sec(T)+v_0 \end{pmatrix}$$

Its implicit equation is:
$$\frac{(Y-u_0)^2}{a^2} - \frac{(X-v_0)^2}{b^2} = 1$$

The asymptotes of the hyperbola have slopes $\dfrac{b}{a}$ and $-\dfrac{v}{a}$.

Student Worksheets

Student worksheets follow.

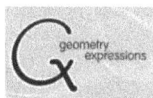

Inside-out Ellipses

In the last lesson, we turned an ellipse inside-out and ended up with a hyperbola. In this lesson, we are going to find general equations for hyperbolas.

Make sure Geometry Expressions is in Radians before you proceed.

1. Recall the parametric and implicit formulas for the Unit Circle.
   Open Geometry Expressions.
   Create a circle with center at the origin, and with radius 1.
   Calculate the symbolic parametric equation for the circle.
   Calculate the symbolic implicit equation for the circle.

Equations for the unit circle

   Now, substitute the parametric equations into the implicit equation. What identity did you create?

2. What do you suppose the graph of $x^2 - y^2 = 1$ would look like?
   First, we need to find an identity in the form
   <trig function>^{2} − <trig function>^{2} = 1.

   Start with the identity you recalled in part one.
   Subtract $\sin^2(T)$ from both sides.
   Divide each term by $\cos^2(T)$
   Compare your result to the pattern: $x^2 - y^2 = 1$
   What is the trig expression that is standing in for x ?

   What is the trig expression that is standing in for y?

   Note: for the next step, if you need to enter

   sec(x), use 1/cos(x)

   csc(x), use 1/sin(x)

   cot(x), use 1/tan(x)

Deriving the unit hyperbola

Draw a function in Geometry Expressions

Select Parametric type

Type your trig expression in for *x* and *y*.

Your result should appear to be a hyperbola. But is it?

The asymptotes of the "hyperbola" that you just
created are the lines *y* = x and *y* = -x.
Add them to your drawing

Create a line.
Select the line, and click on Constrain Implicit Equation

Type in *y* = x

Repeat for *y* = -x

3. Remember that a hyperbola is the locus of points such
    that the difference of the distances from the foci is a
    constant. To verify whether our "hyperbola" is actually
    a hyperbola, first we need to find its foci.

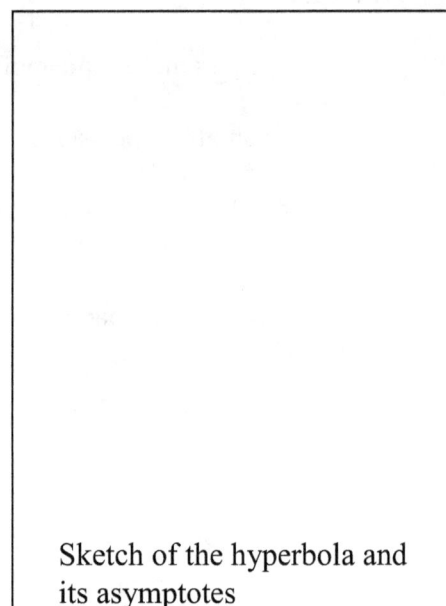

Sketch of the hyperbola and
its asymptotes

If you don't have the unit circle on the same drawing as your "hyperbola," construct it now.

Construct the points of intersection for the circle and each asymptote.
    Select the circle and Select the line

    Click on Construct Intersection

Construct perpendiculars at each intersection point.
    Select the line and the point.

    Click on Construct Perpendicular.

The points where the perpendiculars intersect the *x*-axis are the foci of the hyperbola.
Construct points at those intersections
    Select the perpendicular and the *x*-axis.
    Click on Construct Intersection.

Clean up your drawing by hiding all of the straight lines.
    Click on the line
    Right click
    Select Hide from the pop-up menu.

4. We've found the foci, so now all we need to do is find the constant difference. Use diagram 1 to help you find it

A and D are foci. Consider point C on the hyperbola.

How far is it from C to D (length CD)?

How far is it from A to C (length AC)?

What is AC – CD?

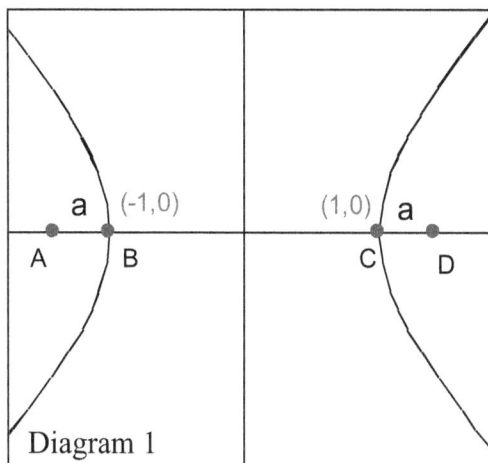
Diagram 1

On your Geometry Expressions drawing, draw a point that is not on the hyperbola or on the circle.
Constrain the distance from one of the foci to your point to be *d*.

What did you get for AC – CD in Diagram 1? Add that value to *d* and use it to constrain the distance from the other focus to your point.

What happens to your point?

Hide the hyperbola.

Create the locus for your point, using *d* as a parameter. Adjust the start and end values if you think it's necessary (you can change them later by double-clicking on the locus curve).

Find the symbolic implicit equation of the hyperbola, and record it here:

5. So far we have found the parametric and implicit equations of the unit hyperbola. But what are the general equations?

Start with a new Geometry Expressions drawing.
Create a unit hyperbola using parametric equations.

Now, modify the parametric equations like this:

$$\left(\begin{array}{l} X = a * \dfrac{1}{\cos(T)} \\ Y = b * \tan(T) \end{array} \right)$$

Change the value of a with the slider bar on the Variable Tool Panel. What happens?

Change the value of **b** with the slider bar on the Variable Tool Panel. What happens?

Create the ellipse with this parametric equation:

$$\left(\begin{array}{l} X = a * \cos(T) \\ Y = b * \sin(T) \end{array} \right)$$

How are the hyperbola and the ellipse related?

Create a line and constrain its implicit equation to be $y = b/a * x$. This line is one of the asymptotes of the hyperbola. What do you think is the equation of the other asymptote?

Test your conjecture on your Geometry Expressions drawing until you get it right.

6. How does a translation affect the equation of the hyperbola?

Delete or hide the ellipse and the asymptotes from your drawing.

To translate the hyperbola:
   Create a vector and constrain it to the default settings.
   Select the hyperbola.
   Click on the Construct Translation tool.

Click on the vector to translate the hyperbola.

Select the new hyperbola
Click on Calculate Symbolic Parametric equation and record the results.

7. Use inductive reasoning to surmise the general implicit equation of a hyperbola, and complete the table

Unit Circle	Unit Hyperbola
$\begin{pmatrix} X = \cos(T) \\ Y = \sin(T) \end{pmatrix}$ $x^2 + y^2 = 1$	$\begin{pmatrix} X = \sec(T) \\ Y = \tan(T) \end{pmatrix}$ $x^2 - y^2 = 1$
General Ellipse	General Hyperbola
$\begin{pmatrix} X = a\cos(T) + u_0 \\ Y = b\sin(T) + v_0 \end{pmatrix}$ $\dfrac{(X-u_0)^2}{a^2} + \dfrac{(Y-v_0)^2}{b^2} = 1$	$\begin{pmatrix} X = a\sec(T) + u_0 \\ Y = b\tan(T) + v_0 \end{pmatrix}$

In the general equation of a hyperbola:

a results in

_____

b results in

_____

u_0 results in

_____

v_0 results in

_____

8. Subtraction is not commutative, so what is the graph of $y^2 - x^2 = 1$?

Solve the equation for *y*.
Open a new Geometry Express file and graph the result. Remember to type sqrt for square root, and to put the radicand in parenthesis.

Is the entire graph shown, or is some missing?

How is this hyperbola different from the others?

Solve for y
$y^2 - x^2 = 1$

We got this graph by exchanging *x* and *y*. Create a hyperbola with parametric equations, but exchange the *x* and *y*. What is the result?

9. Summary

A hyperbola is the locus of points _____

_____

If a hyperbola opens to the left and to the right:

Its parametric equation is:

Its implicit equation is:

If a hyperbola opens up and down:

Its parametric equation is:

Its implicit equation is:

The asymptotes of the hyperbola have slopes _____ and _____.

Lesson 6: Eccentricity

Learning Objectives

So far, the unit has neglected parabolas. In this lesson, we begin with the definition of a parabola as the locus of points equidistant from a point (the focus) and a line (the directrix). Next, the definition is changed to include *eccentricity,* the ratio of the two distances, resulting in new definitions for ellipses and hyperbolas.

Math Objectives

- Define the parabola as the locus of points equidistant from a point (the focus) and a line (the directrix).

- Introduce the concept of *eccentricity.*

- Define ellipses and hyperbolas in terms of focus, directrix, and eccentricity.

- Find the limit of the locus as *e* goes to 0, and compare with the limit of the equation as *e* goes to 0.

Technology Objectives

- Use Geometry Expressions to create a locus of points and derive its equation.

Math Prerequisites

- Knowledge of conic sections presented earlier in this unit.

- Ratios

- Equations of parabolas

- Factoring by grouping

Technology Prerequisites

- Geometry Expressions, as learned in this unit.

Materials

- Computers with Geometry Expressions.

Overview for the Teacher

1. The geometric definition of a parabola is the locus of points equidistant from a line (the directrix) and a point (the focus). Diagram 1 shows sample student work.

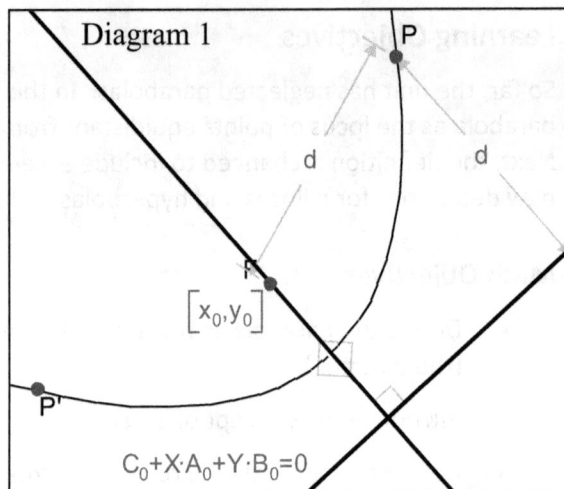

Diagram 1

$$C_0 + X \cdot A_0 + Y \cdot B_0 = 0$$

2. Some students will try to generate the symbolic implicit equation for the curve before nailing down some of the constraints. Make sure they follow the steps to lead them to a simpler result.

   Diagram 2 shows the parabola with the x-axis for the directrix, and the focus on the y-axis.

3. In part 3, the drawing will be essentially the same, but the equation will change to

$$X^2 - 2Yc + c^2 + Y^2\left(1 - e^2\right) = 0$$

   The process of substituting 1 for e will help students see the role that e plays in the equation as well as in the diagram.

   Caution: Some students may think that that e represents Euler's constant. It does not.

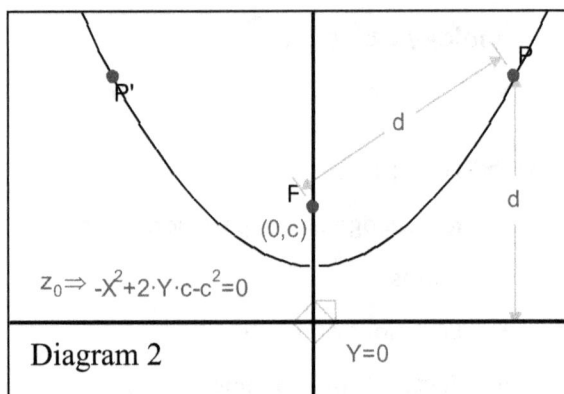

$$z_0 \Rightarrow -X^2 + 2 \cdot Y \cdot c - c^2 = 0$$

Diagram 2 $Y = 0$

Diagram 3 shows a curve with eccentricity less than 1: an ellipse.

The X^2 and Y^2 coefficients will have the same sign in this case.

Diagram 4 shows a curve with eccentricity greater than 1, a hyperbola. Some students may think they have returned to a parabola – help them see that the Y^2 term in the equation contradicts this.

For a hyperbola, the X^2 and Y^2 terms have opposite signs. This corresponds with the general equations they have already discovered.

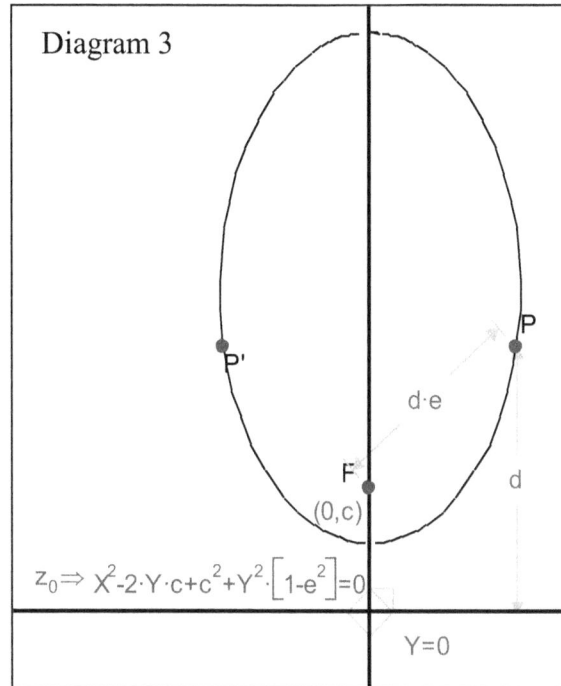

Diagram 3

$$z_0 \Rightarrow X^2 - 2 \cdot Y \cdot c + c^2 + Y^2 \cdot \left[1 - e^2\right] = 0$$

$Y=0$

4. The focus-directrix model breaks down for $e = 0$; but the equation generated resembles a circle, but it is a circle with radius 0.

5. Conclusion:

If $e > 1$, then X^2 and Y^2 coefficients have opposite signs and the curve is a <u>hyperbola.</u>

If $e = 1$, then Y^2 term is eliminated and the curve is a <u>parabola.</u>

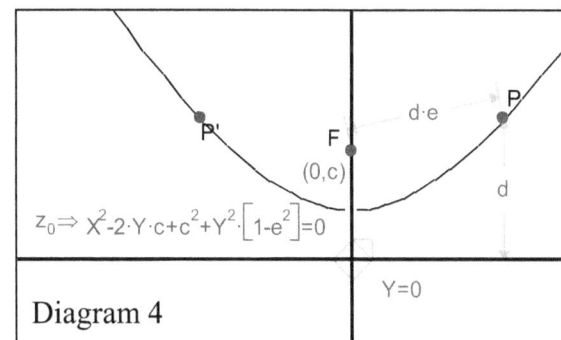

$$z_0 \Rightarrow X^2 - 2 \cdot Y \cdot c + c^2 + Y^2 \cdot \left[1 - e^2\right] = 0$$

$Y=0$

Diagram 4

If $0 < e < 1$, then X^2 and Y^2 coefficients have the same sign and the curve is an <u>ellipse.</u>

As e approaches 0, the X^2 and Y^2 coefficients become the same the curve becomes closer and closer to a <u>circle</u>.

Student Worksheets

Student worksheets follow.

Eccentricity

The final locus that we will look at in this unit is the locus of points equidistant from a fixed line and a fixed point. From there, we'll see what happens if the ratio of those distances is something other than one.

1. What is the locus of points equidistant from a fixed line and a fixed point?

   Open a new Geometry Expressions file.
   Create a line, and constrain its implicit equation to the default.
      Select the line

      Click on Constrain Implicit Function
   This line is called "the directrix."

   Create a point that isn't on the line, and name it F.
   Constrain its coordinates to the default.
   This point is called "the focus."

   Create another point that isn't on the line. Label it P
   Constrain the distance from P to F to be *d*.
   Constrain the distance from P to the directrix to also be *d*.

   Create the locus of point *P* with *d* as the parameter. Set the start value to 0 and the end value to 20.

   To see the whole curve
      Construct a perpendicular from F to the directrix.
      Select the locus.
      Reflect it across the perpendicular.

   What type of curve do you think this is?

2. Calculate the Symbolic implicit equation for the locus

   To make this simpler, we are first going to rotate and translate the locus curve. Remember that rotation and translation preserve the size and shape of a figure.

   Constrain the implicit equation of the directrix to $y = 0$.
   Constrain the coordinates of F to (0, c).
   Solve the equation for y

Do you recognize the form of the equation? What type of curve is this?

Does your answer agree with your guess in part 1?

3. What if the two distances you used to create the locus were different?

    Change the distance from P to F to be *d*e*. The value *e*, which is the ratio of the two distances, is called "the eccentricity."

    Set the value of *e* to be 1, and you have the same curve as in part 2.

    Write down the implicit equation for the locus.

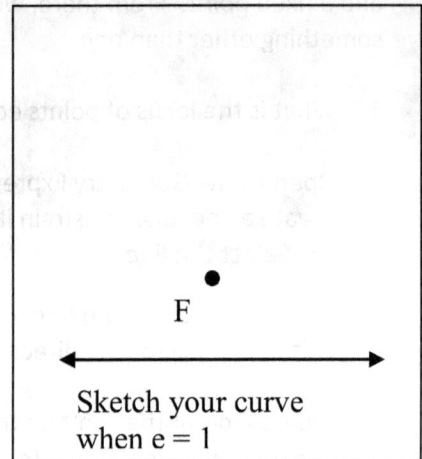

    F

    Sketch your curve when e = 1

    Substitute *e* = 1.

    What is the result?

    If e = 1, then the curve is _____

    Use the Variable Tool panel to change the eccentricity to a value that is **less** than 1.

    Remember that the general implicit form for an ellipse is
    $$\frac{(x-x_0)^2}{a^2} + \frac{(y-y_0)^2}{b^2} = 1$$
    and the general implicit form for a hyperbola is
    $$\frac{(x-x_0)^2}{a^2} - \frac{(y-y_0)^2}{b^2} = 1$$

    Is the Y^2 coefficient positive or negative?

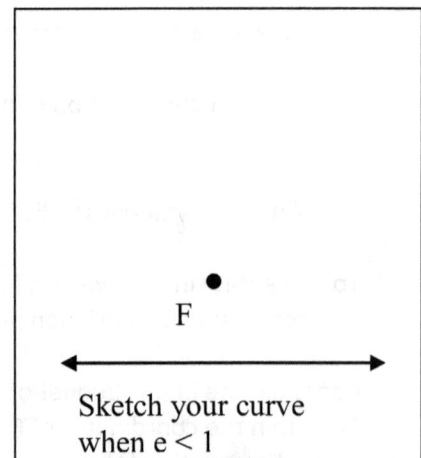

    F

    Sketch your curve when e < 1

Based on this, what type of curve do you think this is? Why?

If e < 1, then the curve is _____

Now, set the value of *e* to be **greater** than 1.

Is the Y^2 coefficient positive or negative?

What type of curve do you think this is?

If e > 0, then the curve is

_____

F

Sketch your curve
when e > 1

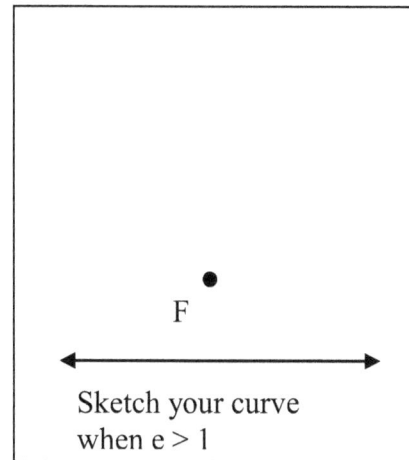

4.   What curve do you expect if the eccentricity is equal to 0?

Use the slider bar on the Variable Tool Panel to move *e* towards 0. If your curve disappears, make values for d and c closer together.

As *e* gets closer to 0, what happens to PF?

As *e* gets closer to 0, what happens to the shape of the curve?

Substitute 0 in for *e* in the equation. $X^2 - 2Yc + c^2 + Y^2(1 - e^2) = 0$

Put the equation into $(x - u_0)^2 + (y - v_0)^2 = r^2$ form.

What is the center of the circle?

What is the radius of the circle?

As the eccentricity gets closer to circle, the radius also gets closer to zero. Therefore the eccentricity model cannot be used to define a circle. It only defined ellipses, hyperbolas, and parabolas.

5. Conclusion

The fixed line is called *the directrix*.

The fixed point is called the *focus*.

The curves we studied are the locus of points for which the distance from the directrix and the distance from the focus had the same ratio. That ratio is called the *eccentricity, e.*

The curves are all of the form $X^2 - 2Yc + c^2 + Y^2\left(1 - e^2\right) = 0$

If $e > 1$, then X^2 and Y^2 coefficients have opposite signs and the curve is a

_____.

If $e = 1$, then the Y^2 term is eliminated and the curve is a

_____.

If $0 < e < 1$, then X^2 and Y^2 coefficients have the same sign and the curve is an

_____.

As e approaches 0, the X^2 and Y^2 coefficients become the same the curve becomes closer and closer to _____.